Ezeani is no ordinary child.
He sees things others don't.

Despite the burden of these visions, his precocious nature blossoms into genius and Ezeani grows up to be a gifted mathematician and physicist. When he leaves Nigeria and his adoring family behind to study at Cornell in the US, he remains haunted by his most persistent vision, Anyanwu, the Sun God. While Ezeani is adjusting to his new life in America, Anyanwu's presence takes on an increasingly sinister and malevolent form – and chaos reigns. It's enough to make anyone lose their grip on reality.

The Comfort of Distant Stars is a bold coming-of-age tale blending physics, philosophy and Igbo cosmology, examining how we understand our place in the universe. It ponders the big questions we all ask ourselves about the nature of time and of being – ultimately revealing the startling vulnerability of the human mind.

Also by I.O. Echeruo

Expert in All Styles

THE
COMFORT
OF DISTANT
STARS

I.O. Echeruo

CANONGATE

First published in Great Britain in 2026
by Canongate Books Ltd, 14 High Street, Edinburgh EH1 1TE

canongate.co.uk

1

British Library Cataloguing-in-Publication Data
A catalogue record for this book is available on
request from the British Library

ISBN 978 1 83726 333 2

Typeset in Perpetua by Palimpsest Book Production Ltd,
Falkirk, Stirlingshire

Printed and bound by CPI Group (UK) Ltd, Croydon CR0 4YY

The manufacturer's authorised representative in the EU for product safety is
Authorised Rep Compliance Ltd, 71 Lower Baggot Street, Dublin D02 P593 Ireland
(arccompliance.com)

MIX
Paper | Supporting
responsible forestry
FSC® C013604

In Loving Memory
of My Brother:
Okechukwu Chima Echeruo
5 April 1971, Ibadan, Western State, Federal Republic of Nigeria
2 July 2015, Kent, Washington State, United States of America

People like us, who believe in physics, know that the distinction between past, present and future is only a stubbornly persistent illusion.

Albert Einstein

Time by itself means nothing, no matter how fast it moves, unless we give it something to carry for us; something we value.

Ama Ata Aidoo

There is no story that is not true.

Chinua Achebe

CONTENTS

PART I

1 The Importance of Eggs   3
2 If You Want to Know Me   19
3 The Appearance of the Maid   32
4 The Distinguished Professor   52
5 The Time I Still Had Friends   68
6 Fluid Dynamics   88

PART II

7 The Migrant Bird   115
8 Black Body Radiation   137
9 Sun Gods and White Girls   154
10 Our Friend Jesus   171
11 In Ordinary Time   191
12 Perpetual Perishing Presents   209

PART III

13 The Varieties of Marital Experience   223
14 Relativity and the Problem of Space   241
15 The Copenhagen Interpretation   253
16 The Speed of Causality (or the Permeability and
      Permittivity of Space)   267
17 Decoherence Time   283
18 The Burial of the Dead   298

PART I

PART I

1

THE IMPORTANCE OF EGGS

The Clearing

One cannot understand life without understanding time. Perhaps that is why the truly insightful child is eager to know which comes first – the chicken or the egg.

When I was barely three years old, I was plucked from my bed, taken to a clearing in a dense, lush forest and made a high chief. The title conferred upon me was the greatest, a title that would take an accomplished, brave, virtuous man a lifetime to obtain.

The rain came down in a cascade of pale beads that broke as they stung my skin and forced me to squint and then close my eyes. I could feel Anyanwu's rough palms grab my hand, dragging me in the direction of his steps. The red, clayey soil, caught between my toes, made me feel that I would slip. I could barely make out shapes in the half-light of the storm. Everything was green and slicked with rain. The air smelt like burnt iron. I felt tired. But I responded to Anyanwu's pull.

There were others on the path. They appeared to be following me. I could hear the voices. But their words were drowned by the kernels of rain falling hard on the leaves and the soft echo of that harsh sound off the wet, green foliage. The voices rose and fell, and I imagined that they were arguing among themselves. In one instant, as the voices dropped, I came to believe the argument was over how much money they would get when they sold me and the share each would receive. I was frightened. The thought had come to me clear and complete, in a manner that I would later learn to refer to as an epiphany.

My first instinct was to share this suspicion with Anyanwu. However, I immediately realised that, if they were going to sell me, Anyanwu would surely be in the conspiracy, since he was the one holding my hand and leading the way. While I was still digesting this thought, the voice patterns behind me changed and I heard what sounded like a woman's voice rising and then falling. I was seized with the idea that the woman was telling the others that I was hungry and needed to be fed. And her concern was sincere, not to fatten me up for sale or slaughter.

Decades later, when I shared the story of this dawn with my wife, she would say the thing she found most remarkable was that at just three years old I had somehow internalised the concept that people could sell others. I reflected on her comment, but I could not understand why anyone would think this the most notable part of the story. This conversation took place in that season when she was taking hormone shots that would enable a technician to harvest eggs from her follicles, a necessary step in the production of the child we were both claiming we wanted. 'Well, context is important. It's West Africa after all,' I responded. 'For literally two centuries, high commerce consisted of kidnapping your neighbours and selling them off. Some of that must have still been lingering in the air.' She would smile weakly and look away. Eighteen months after this conversation our only child, Njoku, conceived of a donated egg, would be delivered in a clean, white hospital room in New York City.

*

4

As if commanded, the rain abruptly stopped and the sun emerged from the clouds, its yellow rays badgering my half-closed eyelids. I opened my eyes. We had come into a clearing in the forest. In the centre stood the trunk of a huge Orji tree, sawed off at about the height of a man's stomach and scraped and sanded till it was flat. I heard the trailing, arguing party enter the clearing and immediately stop speaking. As if they had had a sudden, collective, epiphany.

The tree trunk was alive. Even though the branches and leaves were gone, the roots reached into the earth and large mushrooms were growing low on its side. I was disturbed by the incongruous juxtaposition – that a thing could be alive when all its critical functions had been cut off. I stared, wondering how the tree reconciled itself to living when it had become a table. I stretched out my hand and touched one of the trunk's large roots.

The people in the clearing started speaking again. Most of them were men and women I recognised. At their centre was my mother. She walked up and knelt to wrap her arms around me. Her perfume and its fragrance of fixed flowers filled my nose and comforted me. 'How did you know this would be here?' she asked. And it was then, following her hand, I noticed – sitting on the burnished trunk – white eggs in a clay bowl and a trussed cockerel.

I did not understand my mother's question, nor did I have a clear sense of where I was or what I was doing. Before I could tell her, seventeen elderly men entered the clearing. I counted each one. Unlike the members of the first party, who were relatives and neighbours, these men were completely unknown to me. They wore identical checkered black, red and white cloth tied around their bodies with the loose end draped over one shoulder. Each had a bracelet of red beads wound around one wrist. Their feet were bare, like mine. No one remarked on their appearance. I looked around for Anyanwu. I didn't see him.

A tall man with a greying beard and a hard, handsome face cleared his throat. 'Today, we will make him a titled man.' He raised his hand and pointed at me. 'There were many who doubted. I will add myself to that count. But now, the child – by himself – has brought us all to this place.'

For the first time he looked at me. He smiled, then looked away as his face hardened. 'Those that are still asking: "How can we confer the title Ezeani, the highest Ozo title, on a child that has just stopped sucking on his mother's breasts?" Let Odukwe the Diviner speak. Let no one say they do not know why we, the Nze na Ozo, did this thing.' He paused, pulled at his checkered cloth and threw it over his right shoulder.

The speaker's gaze fell on a man with a bad eye – a pale blue disk floating in a white, cloudy fluid – and a large, bushy beard. The man's clothes were dirty and tattered rags, and a large goatskin bag hung from his shoulders. His good eye darted from side to side as if he was sizing up the gathering. I concluded he was Odukwe, the Diviner.

'This woman's child will not sleep at night.' Odukwe raised his arm and pointed at my mother. 'She says he wanders through the house muttering strange things. And from birth no month has gone by when he hasn't fallen so sick that she has feared his death. Many of you here have witnessed it. They have taken him to the doctors that hang things around their necks at the hospital, but no one has been able to tell them what is wrong with the child.' The Diviner paused. His face was fixed on my mother. My mother started to sob quietly.

'She told me,' he said, turning and pointing at me, 'that this child told her, with his own mouth, that he will die if Umudim does not honour him for the greatness he has conferred on it.

6

When she asked him why, the small child you see before you scolded her and said, "Do you not know how greatly I have honoured Umudim by being born here?"'

Suddenly, abruptly, the Diviner sat on the ground and straightened his legs. He reached into his bag and brought out a carved wooden circle and a piece of chalk. 'She brought him to me, and I have consulted. And it is true. I asked the lizard, and the lizard asked me to ask the nza bird; I asked the bird, and it whispered to me that Anyanwu, the Sun God himself, was eager that this small boy should die so he and the child can converse freely in the spirit world. It was this whispering which made me know immediately that we must convince the child to stay by conferring on him an Ozo title. And not just any Ozo title, but the Ezeani. Why are we to give to a mere child a title which we don't confer on even the most accomplished man? Because Anyanwu knows all things and the child knows something. Let him stay with his kinsmen and perhaps we too will know some of what he knows.' As the Diviner spoke, he drew a horizontal line across the circle with the chalk. Then he drew a vertical line that bisected it in the middle. Then various intersecting diagonal lines, until the lines looked like a starburst.

As he spoke and drew, I stared at him. His story was vivid, but I couldn't reconcile anything he said with the sense I had of myself. I certainly had no memory of any of these events. I had no memory of ever having seen the Diviner.

My first memory is of that walk in a thunderstorm to a clearing in the forest where I was acclaimed a titled lord of Umudim. An egg was broken on my head, the yolk and albumen running down my face – that came first. Then the cockerel's neck was sliced; spurts of its blood showered on

7

me and then the half-alive creature was cast aside, its body hopping in spasms until the blood stopped spurting and the mass of wet feathers stayed still. A few words were spoken, and the seventeen men bent their heads in supplication to me. 'Ezeani! Ezeani!' they hailed. When they came to greet me, they were amazed that I knew all their names and called each by his highest title.

Most people call me Ezeani. They do not realise they have addressed me by my title and not my name. This has happened for practical and easily inferred reasons. My mother and siblings were compelled to refer to me by this title and not my name. A titled man could not be addressed in any other way by any he exceeded in rank. Everyone else, hearing my family refer to me in this way, assumed Ezeani was my name and so addressed me. This of course would have the ancillary, if unintended, effect of ensuring that they too complied with an edict that many of them would have laughed at, especially in application to a child as small as I was then. It is the reason that on my passports and official documents (except my birth certificate) my name is given as Ezeani Kobidi.

I was tired after the ceremony. My mother noticed and put me on her back to carry me on the path back to our home. I wore the red coral bead bracelet that signified my title on my right wrist. One of the seventeen elders could not help chuckling. 'A titled man that rides on his mother's back,' he said, suppressing a smile. But I kept falling asleep and I did not hear or see anyone else's response. Occasionally, I would be awoken by the voices of the party raised in some sort of argument. Once, the voices that woke me were aimed at my mother and me. They sought direction on the best path back to our village. I was irritated by the disturbance and wished only to get back to sleep. 'Why don't you ask Anyanwu?' I

8

muttered and nestled my head deeper against my mother's back.

'Which Anyanwu?' my mother turned to the Diviner to ask. 'The God or a man? Is anyone here called Anyanwu?' The Diviner himself seemed confused, and there was uncertainty in his bad eye whose blue disk focused on me. I felt the bad eye was intent on conveying a slovenly insolence, but I ignored it. The Diviner's good eye darted from one side of his face to the other. He raised his hand and pointed to a path on the right of us. 'I recognise that tree. In a short time, we will be in the village.'

Before he had finished speaking, I had been carried off by the oblivion of sleep. My time stopped. Things happened that I did not see or hear. When I woke, I was lying on a hand-carved recliner chair in our sitting room and my brother Nnamdi was seated beside me, playing with the red Ozo bracelet on my wrist. My sister Obiageli sat on the floor beside the chair making clothes for her dolls with scraps of cloth and safety pins. 'Where did they take you?' she turned to ask me.

I had no clear idea of age then, but I understood that my sister was older than my brother, who was in turn – in some sense that I did not then comprehend – older than me. My brother was still playing with the red coral bead bracelet, sliding it up and down my arm. Suddenly, in a quick movement, he slipped it off my wrist and started to put it on his own. I could sense that he was about to run off and, instinctively, I grabbed the Ozo bracelet at the same time as his arm was moving away. The bracelet, pulled at two ends by forces greater than the tensile strength of the string that held its beads together, disintegrated, and the red coral scattered around the tiled floor. My brother looked stunned. I started shouting and then crying loudly.

The Family Portrait

My earliest memories were of this time when we lived in our village. Before my father's return and the move from the village house in Umudim to the modern bungalow in the university campus at Ibadan. The things I remember, I recall vividly and clearly. But there is much that must have happened then of which I have no memory. My father had left us to go to Brown, a university in Providence, Rhode Island, to get his PhD. I heard all this later. At the time I did not actually understand who he was.

There was a picture in the house in Umudim, framed and positioned on the front wall of our living room, of my father, my mother and my elder brother and sister. My parents were seated, and my sister was standing in between my father's legs with his right hand on her shoulder. My brother was sitting on my mother's lap and her left hand was wrapped around his stomach. My brother had an enormous smile on his face. My mother's smile was only slightly dimmer and there was joy in her eyes. My sister was not smiling but her eyes had a hopeful look. They were focused on the lens of the camera as if from this focus a smile could come. My father – the man who I didn't know – had an enigmatic half-smile, like one who had still not made up his mind.

I looked at this picture often in my many comings and goings from that room and there was a quality about it that disturbed me greatly. One day I voiced my discontent: 'Why am I not in that photo?' I demanded, pointing at the photograph.

'You weren't born yet,' my brother replied, chuckling.

'It's a picture of my whole family, and I am not in it,' I continued, placing my hands on my hips to emphasise my displeasure.

'But my dear Ezeani, you weren't born when the picture was taken,' my mother said. She was smiling quite pleasantly. I must tell you that I could not comprehend the point she was making.

'Why did you take the picture without me?' I asked. 'Where was I when you were taking the photo?'

My sister Obiageli put her hand on my shoulder. 'We couldn't take the picture with you because you weren't born,' she whispered in my ear.

This was perhaps the first instance I can recall of the sense I had, for most of my life actually, of seeing less than others, comprehending less, understanding less. Things seemed to elude me. I couldn't make sense of what seemed obvious to everyone else.

The realisation would come to me much later in life that this was almost the exact opposite of the case. When this realisation arrived, it was a surprise. It was in the three years when I earned my PhD in theoretical physics and published four of my best papers. Those years when the only words that came out of my mouth were 'Yes', 'No' and 'Undetermined'.

'Ezeani, when your father comes back, we will take another photo with you in it. Please don't be angry,' my mother said as she moved close to hug me. She kissed me on my cheeks and her fragrance filled my lungs, and though I was not satisfied, I was pacified and let myself fold further into her embrace.

One cannot understand life without understanding time. Later that afternoon, when my mother and siblings had gone out to the back of the house to gather clothes hung out to dry before the threatening rain would get them wet again, Anyanwu walked through the front door of our living room. He walked straight to the reclined chair where I was sitting and settled next to me. 'Ezeani, I know you and I will be good friends. You ask wicked questions.' He patted me on my knee with his rough palm. He smiled and then reached into a goatskin bag slung over his shoulder and produced a gourd of palm wine, a drinking horn and a few lobes of kola. 'Eat. Drink,' he said and filled my plastic cup with the frothy palm wine. I drank all the sweet wine and Anyanwu refilled my cup. 'Eat some of the kola,' he said, handing me a lobe, 'and empty your cup. It is important to drink. It is important sometimes to be drunk.' I emptied my cup. Anyanwu poured wine into the horn and drank. Then he licked his lips. 'It is a good question! Where were you when they were taking the photo?' and he laughed so hard that his legs started to shake. The palm wine was getting to my head, and I was not sure that what I was seeing before me was true. The room seemed to be floating away from me and I was unsteady as I got on my feet and reached out to touch the receding space.

I sat with Anyanwu, and we conversed. He drank many draughts of palm wine. He filled my small cup again and I drank. He did most of the talking. His head bent back, laughing, as he stroked his chin. Suddenly, he got up. 'Let me go and piss,' he slurred, and staggered through the door. I belched. I waited but he did not return.

And the child learnt something he did not know before – that the world is uncertain; that time shuffles from side to side; that the past and the future appear different only because our vision is blurred; that strange souls live inside us; and that what lives within us is not us.

When my mother returned from the laundry line, I was sprawled drunk on the living room floor. I am not sure of my exact words, but I am aware that I spoke to her rudely with a bawdy familiarity completely inappropriate to our relationship as mother and son. My siblings stood at the door, their arms heaped with dry clothes, and looked on in amazement. My mother could smell the alcohol on my breath and kept asking me, 'Who gave you palm wine to drink? Where did you get palm wine?'

'Why are you asking me?' I sneered, and then with a smile I opened my palm and playfully smacked her buttocks.

My brother Nnamdi burst out in laughter. My mother turned to look at him sternly and then lifted me to her shoulder and carried me to the room I shared with my siblings. I could feel her chest heaving against my cheek and her fragrance in my nose. I fell asleep the moment I felt my head touch the bed.

The next morning, I arose early. My siblings were still asleep, and I quickly bathed myself. I walked to the back of the house where my mother was setting out the breakfast table. I had woken up consumed by a deep affection for her and a desire to do what I could, however small, to make her tasks easier. 'Are you sure you can lift that jar?' she asked, tender, amused and pleased by my enthusiastic helpfulness.

We were eating our bread and laughing when a man came through the door and sat down. He was fat with very yellow skin and a sparse moustache above his upper lip. He laughed a lot. I was able to discern from his conversation with my mother that he was my father's brother.

'Ezeani! Ezeani!' he hailed me. He chuckled. 'I hear you are a titled man.' I looked at him. His mouth was fixed in a wide mocking smile, the lips moist with saliva. He had his hand on his large stomach.

'Ambrose, it is true.' I addressed him with his given name, a name I had just heard my mother use.

He laughed. 'You should call me Uncle Ambrose. Show respect,' he said and then turned to my mother.

'Show respect! You are addressing a titled man,' I shouted and continued eating my bread with my eyes fixed on him. My mother smiled. Ambrose looked back at me. Then he laughed.

Nnamdi and Obiageli were arguing about something and kept calling my attention away from the exchange between Ambrose and my mother. From what I heard, I gathered he thought it would be best if the news of my consecration as Ezeani not be shared with my father. My mother was clear in her rejection of this proposal. 'He left me alone to deal with all these problems. He has no business questioning how I solve them. The child has not been sick a single day since the ceremony and it has now been seven months.'

My father's brother laughed. 'We are not quarrelling,' he appeased.

The House on Campus

I was four years and three months old when the man I would learn was my father returned. My brother was six years and six months, and my sister was a month short of her ninth birthday.

The rain kept falling as the car's wipers beat furiously from side to side. The effort was futile. Sheets of water draped the windscreen, and I couldn't see the road. The car slowed, till it was barely crawling.

My father was driving. He had a big bushy beard that reminded me of the Diviner's and a pair of thick-framed glasses perched on his nose. Behind his glasses, his eyes were small, as if they had been set in a face that was too broad. It had only been a couple of weeks since we had been introduced. I was still watching him. My mother held me in her lap in the front passenger seat. My brother and sister were in the back seat of the car. Nnamdi had spread himself out across most of it, fast asleep, his eyes closed. Obiageli's eyes were wide open. She looked like a nervous bird. Her backbone was pushed against the seat and the left roof pillar. She was fiddling with her blue doll.

I turned to my father who was now leaning forward, peering through the windshield. He used a blue handkerchief to wipe the condensation of our breath from the glass – breath that had gathered and settled because of the temperature differential between the warmth of our family and the cold indifferent storm on the other side. Heat only travels from warm bodies to cold ones, never the other way.

The distance between our village and the university campus at Ibadan is just over five hundred kilometres. The drive, then, in the early 1970s, must have taken about seven hours – if you assume that Umudim stood still, that Ibadan did not move and that it was only our family in my father's car that went on a journey. I had no personal sense of relative motion. I slept and the University of Ibadan was brought and laid to my feet, as if it had been teleported.

I woke to Nnamdi's joyful shouts. When I opened my eyes, he was tugging me by the hand. 'We have come to our new house!' he screamed. Obiageli was excited too. She had a bright smile on her face. As I stepped out of the car, I glanced at my father as he pulled out suitcases from the boot. My mother leaned against the car, talking to him, her mouth lilting with laughter.

We marched through the front door of the university senior staff bungalow in birth order, exclaiming loudly at the large rooms, the furniture made of solid woods, the four-legged radio set. We marvelled at the kitchen of large cabinets panelled in green Formica and a four-burner cooker and oven set. There were three bedrooms, and we happily selected one each. I chose the first bedroom we saw. Nnamdi laughed a sweet laugh at my naivety when he came to, and selected, the master bedroom with its en suite bathroom. Obiageli chose the bedroom at the back of the house, where the light fell in the evening through a window with a large floral curtain and made pretty patterns on the floor.

We wandered outside, through the broad sliding doors at the back of the living room that opened onto a deep covered verandah and then a large yard bounded by a thick shrub hedge at its furthest corners. A slight wind blew. It cooled my skin. At the right corner of the yard stood a tall, majestic Orji tree.

We ran through the yard at a dizzying speed. I tried to keep up with my siblings. When I stumbled and fell, they stopped, came back, and helped me up. 'Are you OK?' they both asked. I did not cry. We continued to run until we were back into the house; we ran through the living room, and through the front door, to the driveway where our father's car was parked. We then turned to our left where a pile of large boulders marked its end. The boulders poked out from the earth like giant eggs, like they had fallen in a cascade from the staggered shelf of large grey rocks. We climbed the boulders and then the shelf of exposed grey rocks with brown veins that smelt of iron. When we reached the height, Nnamdi grabbed my hand and Obiageli's, raising them above our heads like triumphant athletes.

In the shadow of the setting sun my mother made a meal for us. We sat in the dining room beside the kitchen and ate to the sound of jazz flowing in from the radio in the living room. The sound was new to me. I had not heard anything like the sonorous horns that blared and jumped as we ate. My father sometimes snapped his fingers to the music; he looked at us a lot and smiled at us a lot. I watched my mother watching him, a restrained smile on her face. After dinner my father stuffed tobacco into a pipe, flicked a match, and a magical aroma filled the room. I was drawn to that sweet-smelling tobacco in the same way I was drawn to my mother's fragrance. Perhaps then, it is not a coincidence that the years before my father gave up smoking his pipe are the years in which he loved me, and I loved him.

When we were exhausted and ready to sleep, Nnamdi would find out that he had not chosen wisely. He was ordered to remove himself from the master bedroom and return to my room. He walked in and mumbled in irritation, 'This room is my room, and it is also your room. But I am older than you.'

'No! It is my room. I will allow you to stay, but it is *my* room,' I shouted as I marched down the corridor to the master bedroom with Nnamdi trailing me.

'Nne, Nnamdi is saying that it is his room, but it is my room,' I complained loudly to my mother.

'No, I didn't say that,' Nnamdi protested.

'Yes, you did!' I screamed.

'I said it was our room,' Nnamdi claimed.

My parents were both in bed in their nightclothes and my father's right arm was around my mother's shoulders. 'Well, Nnamdi, you have to live with the consequences of your choices,' my father chuckled. 'It is Chief Ezeani's room, but he will share it with you,' he said.

'Since it is my room, I want to sleep close to the wall,' I declared when we returned. Begrudgingly, Nnamdi acquiesced. Then, I was assertive in setting out things that were clear. I had not yet learnt to demur to flawed logic, emotional extortion, threats or the application of force. My brother Nnamdi would soon master me. He would protect me from varied applications of fraud, coercion and intimidation for the rest of my natural life. He would also sometimes exploit me using these same methods.

I would live in that house with my family for five years. We would learn the name of the street where our home stood at the end of a cul-de-sac. We would learn our way to school; learn the shortcuts to the University Staff Club; learn new ways of speaking; and learn of new delights in the company of other children. I would see Anyanwu outside that house – wandering around the Orji tree, sitting on one of the large boulders and taking a pinch of snuff to his nose, and sometimes, trailing me at school, ducking his head shyly behind a wall when he saw me look at him – but in the five years we lived in that home he never came into the house and he never spoke to me. Not until the last year, when Ambrose came to live with us.

2

IF YOU WANT TO KNOW ME

Dead Mice

Warmed bodies, when they get hot enough, give off light. At noon the sun stands directly above us; there is little protection, shadows cower and shrink till there's almost nothing left. I had been watching Nnamdi's shadow. The distance between us seemed to increase the faster I ran, until the shadow disappeared, and I could barely see Nnamdi, his blue-check school uniform a blur in the dust that had been kicked up by all the trampling, running feet. When I finally caught up with him and the others – Ukoli, Gbemi and Chijioke – they had a lizard in a jar and Nnamdi was poking holes in the top of the copper-coloured lid with a rusty nail.

'You have to make enough holes so it can breathe,' Ukoli said.

'Later we can do experiments,' Chijioke added. She adjusted the clasps in her hair.

'It's an agama lizard,' Gbemi said. Everyone looked at him.

'We all know that,' Nnamdi said, turning away from him slowly. Gbemi hung his head.

Nnamdi held up the jar so I could see. The red-headed lizard fixed its gaze on me. It seemed to be making a plea. A wet membrane rose from the bottom of its eyes and snapped to the top, like an inverted blink. The lizard's eyes glistened.

'Is your small brother smart?' Chijioke asked.

'Of course. He is my brother,' Nnamdi said.

'He doesn't look too smart.' Chijioke reached out and took the jar from my brother. 'Is this a male or female lizard?' she asked, pushing the jar before my face. I was silent. The lizard moistened its eye again with the inverted blink.

She sighed. 'He doesn't talk much,' she said, and handed the jar back to Nnamdi.

Chijioke started walking back towards the assembly hall. We all followed. Soon Nnamdi caught up with her. They walked side by side.

Ukoli, Gbemi and Chijioke were all in Nnamdi's class. This was when I was in Class One at the University Staff School. It was the class in which I was supposed to learn to read and write. Try as I may now, I cannot evoke any memories of my classmates. However, I have almost perfect recall of all the seventeen students in Nnamdi's class. I am unable to explain this.

My teacher, Mrs Ajayi, would sit us in a large circle on the floor of our classroom, and arranged in this way, we would be encouraged to read from a basic reader that was passed clock-wise from hand to hand. 'A is for Apple,' a child would yell before passing the reader on to her classmate. Other times, Mrs Ajayi read aloud and we would practise tracing out letters on our individual workbooks. 'B is for Boy,' and my classmates would trace the letter 'B' in both its capitalised and its ordinary form.

After everyone else left to play during breaks, I would sit alone in the empty classroom and stare out the window, looking at the sun, the clouds, watching as the children ran and shouted, and shadows danced on the ground beside them. I was trans-fixed by the movement of these shadows and how they grew and shrank.

Sometimes, herded out to join my classmates for playtime, I would stand beside the merry-go-round, and while my classmates played, look through the louvred windows into our classroom. I would watch the shadows cast on the walls by the planes of light dissected by the louvre blades. It was on one of these occasions that I saw Anyanwu, on the floor with his legs straight out before him, a red, black and white checkered cloth tied around his waist. His chest was bare. On his head was a red, black and white floppy cap on which was inserted a large, erect, white eagle feather.

Anyanwu's face was drawn in concentration, looking at the blackboard and then down at the workbook in his lap. He seemed to be mouthing the alphabet as he drew. I was staring at him presumably trying to learn his ABCs for what must have been minutes before he looked up and saw me. He seemed embarrassed and smiled weakly; almost apologetically, his lips opened to show his white, square teeth. Then he closed the workbook, lifted himself off the floor, and without looking at me again walked out of the classroom. A mistress walked onto the classroom balcony and rang a copper bell, swinging her arms and causing the metal ball to oscillate and collide with the conical sides. Breaktime was over. When I returned to my classroom, I saw Anyanwu's shaky and unsatisfactory tracings of A, B and C in my workbook.

In those early years of primary school, I rarely saw Anyanwu. But I would often discover his unsure and crooked hand in my workbooks. I believe that my uneven performance in those early grades reflected Anyanwu's unwanted and substandard contribution to my own work. It is a conclusion I have reached only in retrospect. Then, I stared out of the window often; I did not really care what was written in my workbook.

I did not grasp in any real way the purpose of school. School and, if I am honest, everything I was asked to do made no sense to me. They seemed arbitrary choices of my parents or teachers. I could not even be sure of this attribution. Even though I received their instructions, I was not certain my parents or teachers were themselves responsible for the directives. I was conscious that they could be agents, passing on edicts they did not themselves comprehend. I certainly only possessed a vague understanding of the fundamental dynamics of a school and did not gather I was being evaluated or graded in any way by Mrs Ajayi or any of my subsequent primary school teachers. It was my father who would bring this assessment activity to my attention later in my primary school career. 'Chief Ezeani, my precocious son. You are failing all your classes!'

After school, we would wait at the lower gate to be picked up by one of our parents. Almost always, this parent was my mother. The Class One rooms were closest to the lower gate, but I was obliged to wait with Mrs Ajayi till Obiageli walked down from the senior classes, a brown leather schoolbag slung over her left shoulder, to collect me.

'Good afternoon, Mrs Ajayi,' Obiageli said.

'Hello, Obiageli, how are you?' Mrs Ajayi responded as she picked up a bottle of fruit juice and placed it into her black bag. I was the only pupil still in class and I could sense that she was eager to leave. Obiageli grabbed my left hand. 'Where is your lunchbox?' she asked.

I released my hand from Obiageli's, walked over to the cubicle and picked up my Superman lunchbox. It was the only one there. Obiageli grabbed my hand again and we walked out the door. Mrs Ajayi locked the door behind us.

At the lower gate, the cars drove in through a crescent, stopping only for the children to hop in before driving out again. When school closed at 2.30 p.m. there would always be a crush of cars. Sometimes there would be a traffic jam and horns were occasionally blared. My mother was always late. By the time she came to pick us up, the traffic had disappeared and there were very few cars. We were supposed to wait for her under an almond tree at the furthest dip of the crescent. Nnamdi ran off to play with his friends.

Obiageli leaned against the tree talking to her friend Margaret. From their conversation I gathered that their attention was focused on two boys playing football on the impromptu pitch across from the almond tree. Obiageli was laughing, one hand placed on my shoulder. A large white Suburban pulled up on the crescent and briefly honked. One boy looked up, reluctantly pulled on his shirt, and walked towards us. He stopped next to Obiageli, leaned and picked up a black schoolbag propped against the tree.

'Hello,' he said to her as he straightened up.

'Hi,' Obiageli said.

The boy smiled and casually walked to the waiting Suburban. Once he was in the car, Obiageli and Margaret dissolved in gales of laughter. My sister looked happy. As the Suburban drove off, its shadow, an absurdly elongated and squished block, made a sweep past the almond tree.

It was my sister who made the decision we should walk home. Neither of our parents had appeared and everyone had slowly dispersed until the pick-up area was empty. Margaret's mother had offered to take us home, but Obiageli said we would wait for one of our parents. Nnamdi argued we should go with Margaret's mother. Obiageli insisted. We waited.

The walk was not a particularly long one. It couldn't have been a distance much more than a kilometre or two. However, we trudged along as if we were on a forced march, our sandals dragging on the ground. I was conscious of our collective sense of abandonment. The future seemed uncertain in ways it had not been before. If we were to be abandoned at school, what other unsought surprises could await us? None of this was spoken. But I sensed it, in myself and radiating from my siblings.

'It's your fault,' Nnamdi muttered several times, looking over at Obiageli.

As we walked past a garbage dump beside the Anatomy Building, we saw a mountain of dead white mice. The bodies piled one on top of another, the animals' white fur and pink feet jumbled up in a stinking heap. The entire building complex had this air, the rotting putrid stench of death laced with a sweet cloying fragrance. I would later learn that this was the smell of formaldehyde, a compound commonly used in biology labs to slow down decay. White whiskers splayed from the pink, still noses. This was my first view of mass slaughter. I felt like screaming. My mouth was open. My brother Nnamdi picked up a stick and prodded the heap. 'They are all dead,' he said.

My sister shrugged. 'Don't touch them,' she warned. 'They could be poisoned or diseased.'

I could not take my eyes off the white hill of dead mice. I stood there and stared; my head swam with visions of scuttling mice and my ears with the sound of despairing squeaks. I felt I was about to fall. From within me, starting at my feet, a silent howl rose like a wave until its soundless vibrations and their echoes bounced around in my head. I would react the same way through my life to the sight of mass slaughter of living beings – whether it was of insects, fish or birds, or much later, in books, pictures of the horrors of the Nazis in Poland or the Belgians in the Congo.

I started running and my siblings, Nnamdi in the lead, ran after me. He seemed to think it was a race.

Sweating, we walked through our front door in birth order and in birth order we saw our mother lying down on the orange cushions of the wooden settee. She had a pillow from her bed under her head. She was sleeping.

Our presence woke her. 'I did not sleep well,' my mother started to say. 'Did you walk back from school?' She looked confused. I imagined her lying in bed beside my father, her eyes open, while he slept. Heat always travels from warm bodies to cold, never the other way. Warm bodies are alive with energy: their atoms vibrate, the electrons jump from place to place. Place a warm body beside a cold one and the warm body will transfer its heat little by little to the cold. The transfer will continue until their temperatures are the same.

Kingsway Stores

My mother pushed the trolley along the supermarket aisle stacked on both sides with tins and boxes. Nnamdi was running ahead pointing to things on the shelves.

'Mummy, please let's buy Ribena.' My mother ignored him, picked up a bottle of Tree Top orange squash and placed it in the trolley. My sister stood by my mother's side. She had placed one hand on the trolley. She seemed almost like my mother's shadow. An image transmuting all her actions, like in a panto-mime.

When we got to the checkout counter my mother let us each pick a treat. 'Don't tell your father that you walked home from school,' she said. Nnamdi nodded eagerly. He understood a bargain had been struck.

An Important Science

It was late afternoon, and the sun was orange, its light diffused through drifting, glowing clouds, bending through the glass panels to frame my father as he smoked his pipe in its rays and their shadows. The air was sweet with the tobacco's fragrance. He started speaking: 'Linguistics is the most fundamental science. Only with the study of language do we truly comprehend the limit of human understanding. In fact, studying language is the only way to understand what it means to be human!'

My father held his pipe in his right hand. His left was draped over the back of the armchair with orange cushions, and he was smiling as he looked around the living room. I sat on his lap; my head erect. Sitting to my left was Professor Ibrahim Addo, a small, bearded mathematician who wore thick glasses that made his eyes appear like enormous balls floating in the middle distance.

'Mathematics is not a language. Not in the way you guys in linguistics think of language,' he said.

Before he had finished speaking, Dr Yemisi, who sat to my right on the long wooden settee – also with orange cushions – interrupted and said: 'I really don't understand mathematics. It is so abstract. In biology we see the link between human life and all life. None of this manufactured exceptionalism that is the stock-in-trade of most of these so-called sciences.'

Dr Yemisi was beautiful. She had a strong nose and oval eyes with light brown irises that were so pretty they frightened me. I kept my eyes on her as she spoke, slowly nodding my head. In her hand was a small glass, half-filled with red wine. A sweet perfume, different from my mother's, wafted from her.

She was about to continue, but before she could speak, Dr Okonkwo, a reader in the Department of English, muttered: 'Humans are different. That is why we have the humanities.'

My father took command of the room with a dismissive wave of his pipe. I adjusted myself on his lap. 'Rousseau says: "We differ from the animal kingdom in two main ways; the use of language and the prohibition of incest",' my father declared. The fragrance of the tobacco cocooned and soothed me.

'That's nonsense. You people in the social sciences should be moved over to the humanities,' Professor Udo said, and laughed. 'There is nothing scientific about what you do!' Professor Udo was a physicist. He was seated next to my father on a leather pouf that barely rose two feet off the floor.

'What do you mean by that? Noam Chomsky's work is not scientific?' my father objected.

'How is it scientific? What experiments did he do?'

'That's nonsense. What experiments did Einstein do? Chomsky is a theoretician setting out hypotheses that can be tested by experiment. Just like your precious Einstein.'

I did not then know who Einstein was. (I had a sense of Chomsky – a figure revered by my father and often mentioned in our household.) I think that these events happened when I was six years and seven or eight months. A couple of years after we had moved to the university house and enrolled at the University Staff School. This was a time when my memories still started and jumped from place to place like a collection of distinct things. The flow of time for me, as captured in my memory, was lumpy. It didn't then have the smooth, deceptive flow that would later mislead me.

At that time, I had received sufficient instruction in mathematics to understand it as the manipulation and extrapolation of numbers. I was therefore intrigued by Professor Addo's implication that it could be analogous to language. Such an idea had never occurred to me. I was afraid that the discussion was in danger of straying from this intriguing idea. 'Please, Professor Addo,' I said, pointing my small index finger at him, 'is mathematics a language? Can you use it to talk to someone?'

'Yes,' he immediately responded. I expected him to continue but he didn't say anything more.

'How?' I asked, looking at him directly, with what I suppose could only be viewed as aggression.

'This, your Ezeani, is something else,' Professor Udo said, glancing at my father. 'He asks good questions. OK, Ibrahim, answer him.'

'It is the only language in which you can say things that are true,' Professor Addo said.

There was silence. Professor Udo smiled.

My mother walked in with a platter of meat pies and Scotch eggs. She was wearing a loose, florid Ankara gown that draped over her body. Her fragrance suffused the room. As she placed the food on the table, she glanced over at my father. 'A true and beneficent wife,' he said. He leaned over and picked up a meat pie and bit into it. 'Please join me,' he encouraged the others, his mouth full.

The Kool-Aid

As I closed the wardrobe door, the darkness was punctured only by a shaft of light that came through the hinges. My eyes were trying to adjust to the sudden blackness. Then beside me there was a sound of rustling clothes. I heard a brief moan. The rustling sound intensified, and I could hear hard, choked breathing. Then a fierce whisper. I tried to look even more keenly through the darkness towards my right, the direction of Chijioke's moan.

Suddenly, the wardrobe door was thrown open and the space flooded with light. My eyes caught Nnamdi quickly taking his hand off Chijioke's blouse.

'What are you people doing?' Gbemi asked, his eyes wide.

Chijioke looked at me, smiled, and then walked past Gbemi towards the kitchen.

'I guess you found us,' Nnamdi said, putting a warm smile on his face. 'Gbemi, do you want to drink Kool-Aid?' He was usually not this friendly with Gbemi. Gbemi smiled and nodded. 'Let's go to the kitchen,' Nnamdi said. I started walking to the kitchen with Nnamdi and Gbemi. Nnamdi stopped and turned, holding my arm to detain me. He waited for Gbemi to pass.

'Why did you come to hide in the wardrobe?' he said in a fierce whisper. 'You caused him to find us!' He pushed against my chest. I didn't say anything. Even though I surmised that the reason Gbemi had known to look in the wardrobe was the sound that Nnamdi and Chijioke had been making.

In the kitchen, Chijioke was stirring the Kool-Aid with a wooden spoon. The motion of the fluid resembled a small tornado, with large and rhythmic crests that reached towards the top of the jug, swirling above a tunnel of progressively lower circuits that formed a rotating funnel. Chijioke lifted the wooden spoon and the funnel slowly started to collapse.

'Come up and get it,' Chijioke said, theatrically tapping on the side of the jug. Gbemi approached with a cup held with both hands. 'First customer,' Chijioke said, and poured the soft drink into Gbemi's cup. Nnamdi had his cup out. 'Beauty before brains,' she said and then poured out the fluid into a cup sitting on the counter, set the jug down and handed the cup to me.

'That's not even remotely true,' Nnamdi said, and chuckled.

29

'You mean he has the brains too?' Chijioke said, smiling.

'You and your sharp mouth,' Nnamdi said. 'Don't deceive poor Ezeani.'

'He looks mighty fine to me,' she said. 'Pity he doesn't talk much. He is what my mother and her colleagues in the life sciences refer to as a "fine specimen".' As she said this, Chijioke held my face between her palms and stared at me. She had eyes with the same pretty light brown irises as her mother, and they frightened me in the same way.

For some reason, the statement or gesture caused Gbemi, Nnamdi and Chijioke to start laughing.

We walked into the dining room and sat at the table drinking the Kool-Aid. Chijioke sat beside me, heat radiating from her body.

Sweet-Smelling Smoke

The pipe had gone out, but my father still held the mouthpiece in his lips. The air was filled with the rich smell of tobacco. We were in my father's study, an alcove built into the corner of the dining room, and separated by a wooden screen door. He was at his desk bent over a typewriter that clattered as he picked at it and made a celebratory ringing flourish every time his left hand slapped a metal handle protruding from its side. I lay on the wooden floor, drawing figures on a large paper sheet. A lamp threw a yellow circle on the wooden desk stacked with books.

Waves of sound, travelling through the air and transferring notes from one molecule to the other, brought a low jazz tune from the player in the living room. I listened for other sounds but could hear nothing. I turned my face back to the sheet and started drawing again. Soon, the sound of the jazz tune and my father's presence seemed to recede and then melt away.

I was startled by the sound of my mother's voice. 'Why is Ezeani still awake? You should be asleep yourself! It is past 1 a.m.' Her head slipped through the screen door. My father turned around, smiling, to look at her.

'My dear, we have made tremendous progress. The book will be finished soon. I even have a title now. Do you want to hear it?' Before my mother could respond, my father picked up a piece of paper from the pile on his left, raised it, and read: '"Assigned Articles: Syntax & Meaning in African Languages."'

My mother smiled. 'That sounds good. It's better than the others,' she said.

'It is going to be revolutionary,' my father said quickly. 'This is epochal work.'

'We pray so!' my mother said.

My father didn't say anything in response. His smile dimmed.

'Ezeani is helping me with cover designs,' he said, turning to me.

I grinned and held up the drawing paper to my mother.

My mother beamed. 'Come,' she said to me. 'Leave the paper on the desk. Say goodnight to your father. It is time for you to sleep.'

I looked over to my father and he smiled and nodded. I rose and hugged him.

He hugged me back, pulling me into his chest. I could feel the heat of his heart through his shirt and the warmth of his arms as they enveloped me. He released me and picked up the paper. 'These are beautiful, abstract depictions of the concepts of the book. This is great work, Chief Ezeani,' and he held his hand out to me. I shook it.

3

THE APPEARANCE OF THE MAID

Massive Objects

Mass is simply the degree to which a body resists movement. Leaning on the door with the weight of my body, I walked into the living room to the sight of Ambrose talking to my mother, a toothpick in his mouth and his head bobbing up and down to nod. His left palm lay on his stomach and his weight compressed the orange cushion of the wooden settee. He was fatter than he had been before.

My mother looked up and stopped speaking when she saw me. Ambrose turned his head. 'Ichie Ezeani!' he saluted me. He was saying the appropriate words, but I sensed mockery in his tone. Still, I nodded in acknowledgement.

'Greet me,' he said, as I stood and silently stared at him. 'I have greeted you.'

'Welcome to our house, Ambrose. No one told me you were coming,' I replied formally. 'Have they offered you kola?'

Ambrose laughed, rolling his left palm around on his stomach. 'Chief Ezeani!' he exclaimed, then turning to my mother he said: 'This, your child Ezeani, is something else.'

My mother smiled. 'Ezeani, your father has gotten Ambrose admission to the Postgraduate School. He will be here for two years to do his Masters.'

'Engineering Masters,' Ambrose interjected.

I was older then. In Class Four of our primary school. My brother was then in Class Six and my sister had moved on to the International School Ibadan, a secondary school affiliated with the university. It was in this class that I had my best teacher. He is the only teacher, other than my first, Mrs Ajayi, that I remember from my time at primary school.

'Come to me,' my mother said, and held out her arms so I could place myself between them and she could close them around me and smother me in her embrace. Her fragrance soothed me, and I held on. Then she smiled at me and said, 'Go to the kitchen; I will come and get you something to eat.'

I walked into the kitchen and moved towards the green refrigerator in the corner. It was mid-afternoon and the sunlight fell in a broad beam on the centre table, throwing a shadow that was shaped like an imploding triangle. As I moved to open the fridge, I screamed, turned, and ran back into the living room.

'There is someone sitting on the floor in the kitchen,' I announced.

'That is Maria,' my mother said. 'She is the maid. She will help us with tasks around the house.'

Ambrose laughed. 'Did she scare you?'

I ignored him and walked back into the kitchen. The girl was still seated on the floor, her feet held straight out before her. She looked up at me when I walked in. Her eyes looked sad, but her lips were attempting a smile. I guessed that she was perhaps a few years older than my sister Obiageli. Her simple dress looked faded and worn. I was afraid she might smell of sweat and dirt, but when I came close, I caught the mentholated scent of medicated soap. I sat next to her on the floor, stretching my legs out straight before me.

'Kedu?' I said. 'Welcome to our home. My mother has told me who you are. I am Ezeani.'

Maria had a large brow, a protruding forehead and thick, bristly eyebrows. She looked at me and then she smiled. Her teeth were white cubes and her face brightened. I could now see she was beautiful. Perhaps even more beautiful than Chijioke.

'How old are you?' she asked me. I told her. We were speaking in Igbo, the language in which I spoke to my mother but not my father.

'I am sixteen years old,' she said.

'I am in Class Four.'

'I have finished primary. I stopped secondary. They told me that after some time in your house they will send me to learn secretarial,' she added.

'What is secretarial?'

'It is where they teach you to type,' she said.

'My father knows how to type.'

'It's not him that will teach me. He is a big man. They will send me to the school,' she said.

'I have a good teacher in my school. His name is Mr Iredua. He says that I am really talented. He says that I will make a great poet or novelist one day.'

'Do you believe him?' Maria asked.

I thought about this question. It wasn't a question that I had asked myself. Before I could formulate a response, my mother walked into the kitchen. She looked quickly at me and Maria on the floor. She shook her head slowly and then smiled. 'Get up, Maria, help me make something for Ezeani to eat.' As Maria rose, I saw that she was tall.

Teacher and Pupil

'Ladies and gentlemen – I'm talking to you, boys and girls,' Mr Iredua said as he dramatically swept at the blackboard with the cloth duster. He shuffled to his chair with a lopsided gait, as if his left foot peered at the ground for a moment before settling on it. 'The poems for this competition must be your original creation. The best three poems will be entered into a contest with those of other boys and girls – I mean ladies and gentlemen – your ages.' There was some laughter. 'Yes, students from all over the world. Do you think we can win?!'

The class erupted in shouts of 'Yes!'

Mr Iredua, our class teacher, led instruction in all our classes, but literature was his favourite. 'Today we are going to read *Julius Caesar*,' he announced as he handed out the Macmillan abridged *Julius Caesar* to the class, shuffling in the rows between our desks.

As he returned to the front of the class, Mr Iredua stopped, turned to me and announced: 'Ezeani, I am expecting a great poem from you!'

Muslims Don't Eat Pork

The blows landed on my back and torso. I was cowering, crunched in a fetal position on the ground as the three boys hit me. Tears and snot made a mess on my face and seemed to provoke even more anger from my attackers. 'Your stupid crying won't save you,' sneered their large-headed leader.

One of them tried to kick me, but missed. I did not protest. I did not understand why these boys had suddenly attacked me as I returned to class after lunch. They took pleasure in it; their smiles were wide.

Someone shouted: 'Stop it!' It was my brother's voice. I saw Nnamdi pull the leader off me and push him into the prefabricated wall of one of the temporary classrooms. Immediately, the other boys stopped punching me, turning away to look at him, their arms hanging by their sides. 'Don't touch my brother or I will beat you!' my brother threatened. I sat up and touched my hand to my lips. The leader had recovered from the push and moved towards my brother. Before he could speak or form a fist, my brother landed a loud slap on his right cheek.

The boy held on to his cheek with both hands. Abruptly – surprisingly – he started to cry. 'He insulted us,' he said, lifting one hand to point at me. I was baffled. I had no recollection of the boy or his friends. 'I'm going to report you,' the large-headed boy said, as he started to run away. His two colleagues followed him.

Nnamdi leaned over me. 'Are you OK?' he asked.

I was still on the ground and my brother was helping me wipe the wetness and mud from my face when the boys that had assaulted me returned with Mr Iredua, two female teachers and a small group of students. The large-headed boy was still holding on to his cheek with one hand while, with the other, he was pointing at me and my brother.

'They were beating my brother,' Nnamdi said before anyone else could speak. 'Why were you beating my brother?' he shouted, looking directly at the large-headed boy.

'Is it true?' Mr Iredua asked. 'Were you beating Ezeani?'

'He said he does not believe in Jesus,' the large-headed boy said.

'And he also said Muslims eat pork,' one of the other boys offered.

36

'Are you Christian or Muslim?' the taller of the female teachers asked me.

'I believe in evolution,' I said.

'But are you a Christian? Do you go to church?' the shorter female teacher asked.

Before I could start a response, Mr Iredua interjected. 'That is why you people were beating him?'

'Muslims don't eat pork!' one of the boys that had been punching me said.

'But Ahmed's father is a Muslim. I saw him eat a pork sausage at the staff club,' I said.

'Is that why you said Muslims eat pork?' the shorter female teacher asked.

Mr Iredua laughed. 'That just means *some* Muslims eat pork. Even if they are not supposed to.'

'If some Muslims eat pork, then what I said is true.'

'No, it is not!' the boy said.

'Muslims don't eat pork,' the shorter female teacher said.

'Nnamdi, did you hit him?' Mr Iredua asked, pointing at the large-headed boy.

'They were hitting and kicking my brother,' Nnamdi responded. 'I just pulled them off,' he lied.

Mr Iredua looked around. 'Everybody, go to your class,' he ordered. 'That includes you, Nnamdi and Ezeani.' He motioned to the boys that had been beating me. 'Come with me. You will be getting detention. You are not supposed to hit people.'

The Fat Ogre

I was flipping through one of my father's books in the kitchen as they walked in. 'Not at all. They are just stupid,' Obiageli said as she turned to Margaret. She picked up a glass jug and poured out a few measures of orange squash and then filled the jug with cold water, stirring with a wooden spoon. 'They always want to pretend they don't care what Kemi thinks. But they always do what she asks.'

'That's what I think too,' Margaret said. 'But Dare likes you. I don't think he likes Kemi at all.'

'I would certainly hope so!' Obiageli said in a singing voice, and she and Margaret immediately started laughing.

'Who is that man?' Margaret asked.

'Which man?'

'The fat man in your living room. He was sitting on the settee.'

'*Mon oncle*,' Obiageli said in an exaggerated French accent.

'*Ton oncle est très gros*,' Margaret retorted, and the girls started laughing loudly.

Mass is simply the degree to which an object resists movement. Heavy things travel more slowly through time.

Ambrose opened the kitchen door. Obiageli and Margaret abruptly stopped laughing. Small defiant smiles still played on the corners of their mouths. They each looked down at the glass of orange squash they had in hand. Slowly, moving his weight along, Ambrose approached the kitchen table where I sat reading. He had a large smile on his face. He pulled out a chair and sat down. Then he looked up at Obiageli and smiled wider. 'What were you guys laughing about? Won't you introduce me to your friend?'

'Friend, uncle. Uncle, friend,' Obiageli said. She and Margaret started walking out of the kitchen, giggling loudly.

Ambrose chuckled. 'Small girls,' he said when they had passed through the door. 'Fine, fine small girls.' He was talking to himself. When I brought up my head to look at him, he muttered, 'What are you reading?' But he was already getting up and leaving the kitchen before I had decided not to answer.

The Plagiarist

It was raining. The raindrops made a dull sound on the roof of our prefabricated classroom. Mr Iredua walked in. His gait caused his head to bop up and down as if he were being carried forward in a wave. We were all seated for the first lesson and the class was quiet in anticipation. He stopped behind his desk and dropped a sheaf of papers. 'These students have written the best poems,' Mr Iredua announced. Then he lifted a single white sheet and read three names. Mine was not one of them.

'They will now come to the front of the class and read aloud for all of us to appreciate. Before they do that, let me congratulate and thank all of you that submitted. Some of them were very good, some of them were not so good, but you made an honest effort, so clap for yourselves.' There were some small claps. 'However, some of you disappointed me by copying other people's poems and pretending they were your own. Did you think you would trick Mr Iredua? There were four of you who did this. I am very, very disappointed. Remember, class: "Crime does not pay!" What did I say?!'

The class, in unison, shouted back, 'Crime does not pay!'

The students whose poems were selected stood in front of the class and read their compositions. I was in a state of both agitation and anticipation as I listened. I recall my bitterness as I heard them read poems far inferior to 'The Comfort of Distant Stars', the poem I had submitted. In any of my previous classes, I wouldn't have had any expectations. Probably, if I am honest, my expectation would have been the exact opposite. I barely registered my other teachers and I suspect I did not register with them either. However, in the six months he had taught me, I had developed such confidence in Mr Iredua, his intellect and his ability to appreciate the work I did. I suppose that it was not a response as analytical as I now recall. It is probably more accurate to say that I felt Mr Iredua sensed my potential and made some effort to nurture and protect me. I was therefore genuinely disappointed that he had not selected my poem. As hurt as I was, I was still unsure of my talents and was hoping that I would learn from Mr Iredua in what ways my poetry had failed to satisfy his requirements. Perhaps, I speculated, there was some criterion, length or subject matter, which I had somehow missed or misunderstood.

This self-doubt would follow me into physics. And in many ways, it would never leave me. When I sent out my first papers, I was certain that the work was seminal, in fact revolutionary – yet, in spite of my supreme confidence in the work I had done, I was simultaneously, in another part of me, utterly uncertain of its merit, expecting to be declared a quack who barely understood the first thing about the science. It appeared to me before publication as if the world's responses to my papers were as full of random possibilities as an application of Schrödinger's wave equation.

In the case of my poem submission, the result was beyond the set of possibilities that I would have included in any wave equation.

'Ezeani, you are a plagiarist!' Mr Iredua declared after class. 'Do you know what that is?'

'Someone that copies the work of others,' I replied.

'Exactly,' Mr Iredua said. 'Why did you do it? I am very disappointed in you.'

'But I didn't copy anyone, Mr Iredua,' I protested.

The class was empty, except for me, Mr Iredua and the three other students he had asked to stay back.

Mr Iredua called the other students one at a time and handed them their papers in turn. 'You copied Soyinka. You copied Robert Frost. You copied Kipling, of all people,' he said.

'Who did I copy?' I asked.

'I haven't figured it out yet, Ezeani. But there is no way that you wrote that poem.'

'I did not copy anyone,' I pleaded.

'All of you, get out of my sight,' Mr Iredua said. 'And don't ever try that nonsense again.'

'But I didn't copy anyone. I wrote it myself, Mr Iredua!' I cried.

'Get out of my sight now!' Mr Iredua shouted.

I walked out of the classroom. Most of the class was gathered in the covered walkway, waiting for the light rain to stop. I walked into the rain and through the field at the back and the thick shrub. I sat on the wet grass at the edge of the drainage ditch that marked the border of the school and I cried.

Manoeuvres in the Dark

'How was Mass?' my mother asked.

'It was very short,' Ambrose answered, then laughed. 'Much shorter than in Umudim. The priest here does not waste time.'

'Yes,' my mother agreed. 'It is the kind of community we have here.'

When Ambrose and the maid came, my mother had made a show of dragging us to Mass for the first two Sundays. This was because Ambrose had said, 'Where are we going for Mass

41

tomorrow?' and my mother always wanted to make everyone feel welcome. My father did not go. Everyone else squeezed into my father's car for the very short drive to the Catholic chapel. But after those first two trips, Ambrose and the maid walked to church on Sundays by themselves.

My mother was turning the stew on the stovetop. Ambrose was fanning himself with a church circular.

'Maria, change from the church clothes and come and help me set the table,' my mother said, turning to catch Maria standing at the corner of the kitchen with a missal in her hand.

'OK, Ma,' Maria said, as she walked through the back door of the kitchen towards the boys' quarters that was hidden from the bungalow by a large shrub.

The boys' quarters had two rooms and a bathroom at the end of an open corridor. Ambrose occupied the larger first room, where he had installed curtains and a television. The smaller bedroom had an empty bed and Maria's clothes. Maria no longer slept there. After her first night in the boys' quarters, she told my mother, 'Evil spirits were banging on the door. Kpa! Kpa! Kpa!' Tears streamed from her eyes. 'I put the bed against the door and prayed to Jesus and the Blessed Virgin to protect me. When I started praying loud and strong it chased the evil spirits away.'

'Are you sure these spirits are not human beings?' my father had said when my mother reported the story to him.

'The neighbours' houseboys wouldn't dare come over at night and knock on our boys' quarters doors,' my mother said. My father grunted. 'Maybe she is not used to sleeping in a room by herself,' my mother mused. My father's face had turned away. He walked through the door, into the living room.

Ever since that first evening, after she cleaned the kitchen and finished her chores each night, Maria laid a mat on the kitchen's linoleum floor and slept facing the green cabinets. Early in the morning, before my mother rose, she folded the mat up and went to the boys' quarters to take her bath and change.

That Sunday afternoon, immediately after Maria walked towards the boys' quarters, I went to look for Nnamdi. He was sitting outside on one of the large egg-shaped rocks at the front of the house.

'What are you doing?' I asked.

'I'm looking for lizards,' he said.

'Why are you always hunting lizards?'

'I don't know.'

'Maria is going to change,' I told him.

'Now!?' he shouted and started running round to the back of the house.

I followed him. When I caught up, he was at the side of the boys' quarters, tracing his finger along the walls.

'Too late,' he muttered. 'She's finished changing.'

As that moment, Maria walked past us towards the main house.

'Nnamdi, kedu?' she called.

'I'm fine,' Nnamdi responded and smiled.

Nnamdi and I stood in the grass and waited for her to enter the kitchen through the back door.

'Let's go and see anyway,' Nnamdi suggested.

A broad walkway separated the boys' quarters' bedrooms from the toilet and bathroom. We had thoroughly explored our new residence when we first moved in. Long before the arrival of Ambrose and the maid, Nnamdi had discovered the peepholes placed in the ceilings of both the boys' quarters' bedrooms. We had not spent much time speculating why someone – presumably someone that had lived there before us – had set up this spying apparatus. The peepholes were accessed by climbing footholds at the back of the walkway that split the bedrooms from the toilet and shower. We climbed up the footholds and lay flat on the platform above the ceiling, looking down into the maid's room. The bed was unslept in, with smoothed bedsheets neatly tucked in. Maria's clothes were in a small bag. There was nothing to see. Nnamdi got bored and we climbed down.

When we got back into the house, everyone was seated for lunch, my father at the head of the table. He was sipping blood-red wine from a large goblet.

'Where have you rascals been?' he said. 'Ezeani, have you joined Nnamdi in his mischief?'

We sat down to eat.

'Pass the rice,' my father said, gesturing at my mother. My mother picked up the bowl of rice and passed it to my father.

'Be careful,' she said. 'It's hot.'

'Ambrose, how is it going? Are you paying attention? This is a huge opportunity for you. Ibadan is academically rigorous.'

'Brother, I am paying attention. It will be well in Jesus's name,' Ambrose replied.

'Jesus won't take your exams for you,' my father responded.

We were still eating rice and stew when Maria walked in carrying the apple crumble my mother had baked. It smelt good. My father looked up and smiled.

After lunch, I wandered outside with my brother Nnamdi. We were looking for lizards.

'I want to catch a big male one. They are the best for experiments,' Nnamdi said.

We chased a few but I did not sense that Nnamdi had his usual thirst for this hunt. He seemed preoccupied.

'I don't know why Uncle Ambrose has a television in his room and we don't have one in the house?'

'Daddy says television leads to feeble minds,' I said.

'But how do we know that is true?' Nnamdi asked, smirking.

Nnamdi was holding my hand, and we were walking in the backyard. We each had a stick which we used to beat the grass and rustle lizards. A few small female lizards ran from us. We did not bother to chase them.

'Let's go to the front of the house. The big lizards like to sun on the rocks,' my brother said.

I followed him and we climbed the rocks. We didn't see any lizards. Nnamdi lay down on a large rock and closed his eyes. I copied him. I could feel the sun on my skin and on my closed eyelids. Soon I heard Nnamdi snoring gently. We continued to lie in the sun.

'Nnamdi, you will get sunstroke. Go and sleep in your room!' my mother exclaimed from the front door. 'Ezeani, come out from under the sun.'

Nnamdi and I got off the large rock and walked into the house. Nnamdi was sleepy and walked to our room. I sat in the living room for a few minutes, watching the birds swooping past the verandah, then I wandered into the kitchen and drank some water. My mother sat at the kitchen table. She was looking through a magazine that had lots of colour pictures. She flipped through the pages, stopping to stare at an image of a woman stepping out of a car with a hand raised high in the air and a smile on her face. My mother's hands were thinner now than they had been in Umudim, and her face was drawn. She frowned more. But she always smiled at me.

'Mummy, where is Maria?' I asked.

'I don't know,' my mother answered distractedly, without looking up from the magazine. I placed the empty water cup in the sink and walked out the kitchen's back door.

45

I wandered down the short path to the boys' quarters. The only sound was from Ambrose's television. It was turned to a high volume. In the foreground, a man was giving descriptions of what players were doing in a football match, and in the background, there were the shouts and chants of a crowd. I walked to the footholds and crawled up to the platform that ran the length of the boys' quarters' bedrooms. I positioned myself above the peephole that looked down on Ambrose's room. I wanted to watch television.

When I peered, I could not see the television. Right below the peephole was Ambrose's bed. His massive frame lay on Obiageli. His hand covered her mouth. He grunted. She whimpered and cried. His fat fingers twitched.

I quickly turned, crawled, and then ran away.

At dinner, my father sat at the head of the table. Obiageli sat to his left. My mother had prepared moi-moi and stew. Nnamdi ate three portions, unwrapping the steamed bean cakes from the green leaves and watching the steam float momentarily over the table before condensing on the far edges of the dinner plate.

'Obiageli, is that all you will eat?' my mother asked. Obiageli didn't answer.

'Where is Ambrose?' my father asked.

'He said he would go and visit some of his friends on campus. He will eat with them,' my mother replied.

Obiageli got up from her chair. 'Have you finished eating?' my mother asked. Obiageli barely nodded her head.

'Don't force food down her throat,' my father said.

Obiageli looked around the table. 'I am going to my room,' she said, facing my mother.

Her face looked like that of a frightened bird.

'Please, another moi-moi,' Nnamdi said, handing his plate to my mother. 'And stew!' My mother chuckled and unwrapped two moi-moi cakes on Nnamdi's plate. As my mother was ladling the red stew, Obiageli walked through the door of her room and closed it behind her.

I woke confused, moonlight falling in a beam through a window onto my face. I did not know where I was. It took me a moment to recall that I had fallen asleep hidden behind the settee in the living room. Then another moment to hear the sound that had probably caused me to wake. It was the sound of my father's grunts coming from the kitchen. I walked over quietly. My father was seated on a chair and Maria, the maid, sat astride him with her mouth wide open. I looked. And then I looked some more. Maria was still beautiful; I could not see my father's face, but his voice was hoarse.

I turned and slowly walked down the hall. My eyes were hot. I did not know what to do. I wanted to go to my mother. To tell her the things I had seen, but mostly so I could crawl between her arms, so she could hold me, so I could smell her fragrance, and feel her comfort. As I reached my parents' room I lifted my eyes to turn the lever on the door. Anyanwu stood before me. His bare feet were shoulder-width apart and his arms were folded on his chest. He was dressed in the same way – a black, white and red checkered cloth wrapped around his waist and then thrown over his left shoulder. A floppy cap, with a giant white eagle feather, was on his head. The muscles of his chest and arms rippled as if he were flexing them. His skin was dark and glistened in the weak light in the hallway. His eyes were fixed on me. I was shocked to see him there. I had never seen him in the house in Ibadan. When I had seen him, he had been outside, far away, avoiding my gaze. I suppressed my instinct to scream.

'Where are you going?' he asked me. I was surprised to hear him speak. I had not heard his deep guttural voice since he had come to our house in Umudim, and we drank palm wine together. That event had always been sharp and distinct in my mind, but I had nevertheless begun to wonder if, because it was so early in childhood, the encounter was one I had somehow imagined, or conflated, into the clean and sharp memory that had stayed with me.

Anyanwu put his arm around my shoulders. Then he turned and we walked into the empty living room. He sat down on the rug in the middle of the floor and beckoned me to sit beside him. The curtains were drawn but some moonlight slipped through where they touched, lighting Anyanwu's hard, handsome face.

'It is time to drink,' he said. He reached into his bag and retrieved a large brown gourd whose mouth was plugged with green leaves. Then he pulled out two ivory-coloured drinking horns. He poured out a full horn and handed it to me. My nose was filled with the fresh, frothy smell of palm wine. He looked at me.

'Ezeani, Ezeani, Ezeani! How many times have I called you?' Anyanwu asked.

'Three times,' I replied.

'Drink,' he commanded.

I lifted the horn to my lips and emptied it. I belched and handed the horn to Anyanwu. He filled it again.

'Drink. Drink. The palm wine will not finish,' Anyanwu said.

When I had drained three horns, I sipped the palm wine slowly. My eyes were no longer hot. 'Why did you stop me from going to my mother?' I asked Anyanwu.

'I did not stop you from anything,' he said and turned to look directly at me. 'Do you know who I am?' he asked.

'You are Anyanwu,' I replied. Anyanwu laughed. He slapped his thighs, then he laughed some more.

'Ezeani, that is true. I am Anyanwu, but who is Anyanwu? But, before you say that Anyanwu is Anyanwu, with your clever mouth, Ezeani, let me tell you.' He turned and looked at me with a small smile. Then he stood up and hit his hand against his chest.

'I am Anyanwu, the Sun God! The Seer of the Universe that sees both what is seen and what is unseen; I am the God of light; the God that coaxes life and lets it thrive . . .'

I started laughing, slapping my thighs. Anyanwu stopped, staring at me in disbelief. Slowly he shook his head from side to side.

'It is not you I blame,' he said sadly. 'It is not you.'

Then Anyanwu sat down again beside me. He poured out a horn of palm wine for himself, then refilled my horn. As he drank, Anyanwu chuckled.

'Do you know why I said I did not stop you from going to your mother?' he asked after a few moments.

'I do not know,' I replied.

'Because I could not stop you,' he said.

'I thought you just said you are a God? Surely a God can stop a child from going into his mother's room,' I responded.

'No. Human beings have mmuo, they are spirits. Just like Gods have mmuo. Nothing with mmuo, this sapience, can be made to do anything against their will.'

'I did not know that,' I said.

'How will you know? No one teaches children anything of value any more. All they talk about in Umudim now is Jesus.' He shook his head.

'Let me ask you,' Anyanwu continued, 'in all the stories of Gods that you have heard, is there any one in which a God made a human do anything? He may ask them to do it; he may threaten to punish them if they don't do it; he may promise to reward them if they do it; he may inflict awful pain on them to compel them to do it; but is there any story that you know when a God just *made* a human being do something?'

I thought of all the stories that I had heard or read: of a man getting swallowed by a fish because he refused to obey an order; of fantastic threats of punishment made to humans to compel them to behave. I could not immediately recall a story in which men were simply made to do what was required of them.

'Can you think of even one?' Anyanwu asked. 'You will not find one. That is because human beings are spirits. And no spirit can make another spirit do anything.' I looked at Anyanwu. He had taught the child I was then something that he did not know.

We drank in silence.

'You are serious in claiming that you are a God?!' I asked. 'You look like a man.'

'Ezeani, what do Gods look like?' Anyanwu asked. He paused, then he said, 'When a human being sees a God, it must be in the form of a human, or if it is not a human, in the form of something humans have seen before.' I looked again at Anyanwu and sipped from my drinking horn.

Then Anyanwu asked me: 'When you saw me in front of your mother's room, why did you not continue to her? Why did you change your mind?'

'You made the path more difficult,' I responded.

'That is true. But you also thought a different thought,' he said.

'That also is true,' I said.

As we drank, we suddenly stopped our talk. I could hear my father's footsteps down the corridor, creaking softly as he returned to the room he shared with my mother at the end of the hall. We heard the soft sound of the door open and then close again.

We sat in silence for a few minutes and then abruptly, as if his mind had been preoccupied, Anyanwu announced: 'Jesus is now among the powerful Gods,' and then cleared his throat.

'What do you mean "now"?' I asked.

'Now, almost everyone worships him. It has made him powerful,' Anyanwu said.

'You are telling me that Gods are powerful based on the people that worship them?' I asked. 'That isn't true.'

'Who told you that?' Anyanwu asked. 'They lied to you. All Gods need worshippers, even if they pretend not to. As human-kind need Gods, Gods need humankind. Why do you think Gods are always making promises and threats to human beings? Why are they angry when you worship other Gods? I used to do the same thing.'

Anyanwu sat silent for a moment, his eyes fixed on me. 'I want to be honest with you. I want you to worship me. Before you say anything, let me tell you what I will do for you . . .'

Anyanwu made wild promises to me. But I barely heard him. As he spoke, I thought of my father. I didn't believe Anyanwu. I gave no credence to his drunk, animated pledges.

4

THE DISTINGUISHED PROFESSOR

Before God and Man

One cannot understand life without understanding time; and one cannot understand time without understanding space. The two are helplessly linked; the faster you move through space, the slower you move through time. If you move through space fast enough, time stops altogether.

The smell of the palm wine was in my nose. As Anyanwu spoke, his face with its strong flaring nostrils seemed to melt away and I saw the face of Maria, the maid, before me. Her eyes were closed, and she seemed to move slowly, a rhythmic motion like a wave rising and falling as she straddled my father. Every movement lingered. I could make out the muscles on her face gradually forming a grimace and then the lines of a scowl. She opened her eyes. She fixed them on me. Then, languidly, she bit her lip. The excruciatingly slow movement confused me. I wondered if anything was happening. I wondered if time had stopped. Anyanwu still appeared to be speaking but the sound that left his mouth felt like it was taking an eternity to get to me. And while I waited, the image of the maid, barely moving above my father, the features and set of her mouth fixed, stretched before my eyes. A deep instinct told me that time had not stopped; I sensed that time was still moving forward, carrying me with it. I could sense the moment associated with each movement as it started, appeared to dissipate, and then gradually built up and bubbled into another moment that also became a movement.

Suddenly, like a movie that had been sped up, the images in front of me scampered up to the usual speed. Anyanwu was speaking and gesticulating wildly. It took me a few moments before I could make sense of his words.

'Children of humankind do not oppress the children of spirits. Children of spirits do not oppress the children of humankind,' he said. He poured some palm wine onto the ground in libation.

'It is the Diviner. He is my only remaining worshipper. He is an old man. He will die soon. If he dies and there is no one else who believes in me, I too will cease to exist. I will die. I, Anyanwu, the Custodian of the Sun, the Keeper of the Firmament of the Stars, I, Anyanwu, the House of Wisdom, I will also die.' He shook his head as if he could barely believe it himself. He continued talking about the Diviner. He seemed to be saying that the Diviner was not reliable. 'It is what happens when a God has only a few worshippers. The God needs them too much. They see him too often. Familiarity and contempt rise to fill the place where reverence ought to be. Insult enters the matter.' I could not follow much of what he said. My mind kept wandering to the maid Maria's eyes. And sometimes her lips.

Anyanwu placed one hand on my shoulder. My mind slowly focused on his face. He was saying: 'The Diviner's bad eye is his good eye. He could see I wanted you as an acolyte. His jealousy blinded him, and he conspired to keep you in the world of human beings away from me. But I, Anyanwu, the All-Seeing-Eye, the Peerless Keeper of Light, I had peered into his head, and I let him deceive himself as to my true intentions.

'Ezeani, I want you to worship me. I need you to worship me. It is hard to explain this. You laughed when I, Anyanwu, the God of the Sun, opened his mouth to speak and beseech you. A voice that once would have made thousands of the wisest, bravest men in the world prostrate in abject terror. That is the voice that you laughed at. But it is not you I blame. A terrible thing happened. And my hand is not without blame.'

He placed his other hand on my shoulder as he turned to face me. I looked into his eyes, and he blew a froth of palm wine and spit on my face. My eyes closed. An icy blast of cold air descended on me. My body started to shiver. A crucifix appeared before my closed eyes. I could see as clearly as if my eyes were open. A pale-skinned man was pinned to the wooden cross. His eyes were open wide with terror. The cross was made of thick wooden beams like those that I had seen spanning walls to hold up roofs. The pale-skinned man's arms were spread and nailed at each end of the beam. His feet were placed together and nailed to the wood. The nails were thick, rusted spikes. Blood, dark red and congealed, was pooled around the wounds. I could see where the blood had dripped onto the cross. The smell was like burnt iron. I could hear a large roar of voices. I turned to look. There were thousands of people. Tens of thousands, stretched out behind me, and I could not see where it ended. The crowd looked like the people I knew from the time we lived in the village in Umudim. They were dressed in the same way, in simple clothes. I could pick out some faces in the crowd. The men, women and children were weeping and crying. Some of the women were wailing. They all pointed at the crucified man. Suddenly, some of the men threw themselves on the ground, as if the pain and agony was too great to bear standing. They rolled in anguish, throwing up small puffs of sand and dust. There was so much pain, and the cries and tears seemed to wash over me. I started to cry.

'Yes, yes,' Anyanwu's voice came. 'This is where all the people that used to worship me have gone. They now worship Jesus. I, the God that Holds Knowledge, I should have seen this. But I didn't. When the other Gods of our Cosmos warned of the danger, in my arrogance, I laughed.'

The crowd was still crying and wailing. It didn't seem they could hear Anyanwu. I looked at the pale-skinned man on the crucifix; he seemed to be listening, his head cocked. The terror seemed to have disappeared from his eyes.

'He can hear me,' Anyanwu confirmed. 'He is laughing at me.' I looked at the man on the cross. He was not laughing. There was a small, wry smile on his face that quickly disappeared. The terror returned to his eyes. The crowd surged forward. The wails and crying now rolled like thunder, from the back of the crowd to the front, where I stood before the crucifix. Anyanwu walked in front of me and placed both arms on my shoulders. Everything else disappeared.

'Ani, Empress Goddess of the Earth, cursed me. "Your foolishness has destroyed us," were the last words she spoke to me.' Anyanwu's hands were still on my shoulders. 'She killed herself. You did not know Gods do that? Everything that has spirit can have its soul broken like a calabash.'

A Convocation in the Sun

These events took place in the week before my father was made a full professor of linguistics at the University of Ibadan. At the convocation, he was inaugurated, and I watched him walk across a stage wearing a shimmering red and black gown over his blue suit. Around his neck was a cape that fell in a V-shaped loop over his back. He wore a funny felt hat that flopped on his head, its gold-coloured tassels swinging against his left ear. My father drew his shoulders back, looking around, his eyes full of pride. I was seated with my mother and siblings on wooden chairs set on a raised platform in a special area for the family of the faculty. My mother beamed at the new professor of linguistics. She clapped. Obiageli was looking up to the stage, her hand cupped above her eyes. Nnamdi was clapping so enthusiastically that a man in a dark suit seated next to us turned to look at him.

At that moment, a short plump man tapped a microphone. My father stood even more erect, and the man started to read a citation of my father's academic achievements. In a flourish, the man ended: 'We, the Faculty of Social Sciences, recommend to the Chancellor and senate of the University of Ibadan that Christian Chikadibia Kobidi be admitted as a professor.'

Between Man and Wife

My wife said to me: 'Ezeani, you have the most imaginative creation story. When you speak about Anyanwu there is a moment where people wonder if they should take you seriously. Before they smile and play along.'

These words were spoken one late winter afternoon in a New England hotel room. I looked out through the windows at the grey sky and the slow dark clouds that clung to the edge of the horizon. I was lying on white bedsheets and my wife was sitting beside me, with her feet on the floor. She wore a cream-coloured cardigan over a light blue blouse. My wife always dressed well. The clothes, I knew, cost money. She placed the pink fingers of her left hand on my chest, patting me distractedly. As she did this, the two rings on her fourth finger hit coldly against my body. I closed my eyes. Pauli's Exclusion Principle holds that two electrons cannot occupy the same space. I allowed myself to think that this was the reason my wife's hand did not pass into my chest and viscera.

Just then the telephone rang, and Heidi picked up the handset of the sturdy black antique resting on the night table. She seemed to listen to someone for a few moments and then she said: 'Let them know Professor Kobidi will be down presently.'

This was in the period, a couple of years after our wedding, when my wife had decided that I was not quite as well known outside academia as the quality of my work demanded. She had taken it upon herself to arrange a programme of speaking engagements at locations in, or close to, university campuses in the continental United States. It was her idea that I should make public reference to Anyanwu. She wrote out a few draft 'treatments'. Her face was bright with the smile of love and hope when she gave them to me.

The pathway from the hotel entrance to the waiting car was a mess of snow, melted ice and salt. The sludge was dirty and tinged with blue. As I pushed my head into the vehicle, my wife asked: 'Should we put some time in the schedule to visit your sister?' Obiageli's face sprang into my mind. It was her face at twenty-two. I hadn't seen her in fifteen years.

'We have a few days in New Haven. She lives less than twenty minutes away,' Heidi said.

I looked at my wife. My face must have seemed helpless. She patted my hand.

'We don't have to decide now. I sent her an invitation to your event. Maybe she will come.'

I settled myself into the back seat. It was warm. I dreaded the moment that I would leave the car and mount a brightly lit stage to provide a ridiculous, simplified summary of my life's work. I especially despised the question-and-answer sessions where inevitably I would suffer through inane questions from people who had made no effort to try to understand what I had said.

'Is there a Q&A?' I asked Heidi.

'No. I insisted that it be eliminated from the format. There are a few pre-submitted questions that will be read at the end. I had one of your graduate students select the most meaningful.'

'Which one of them?' I asked.

'Oh, I can't remember,' she said and waved her hand, as if in exaggerated exasperation. She turned her face from me and looked out the window. The darkness of a New England evening had already descended. My question was which of my graduate students had made the selection, but I was not sure she had understood me. I reached for Heidi's hand.

We drove on in silence.

I did not understand then the true meaning of my work and, deceived by this ignorance, I was only interested in speaking to those who could understand the deep language of mathematics that described the building blocks of the cathedral of our universe.

A Convocation in the Snow

My wife walked ahead of me. It wasn't really cold any more, but as I walked on the short pavement that led to the auditorium's back door, I pulled the coat close to my chest. I could hear the hollow clicking of my wife's heels. When Heidi pulled at the auditorium's door, it opened with a creak and the sound of scratching metal.

The hall was smaller than I had expected. There were only about thirty people in the room. As I sat on the bench beside the stage waiting for the event to start, my wife huddled with a small group of women and men at its foot. I recognised among them Philip Bousquet, an associate professor of physics at Yale University. He wore glasses on an earnest honest face which he revealed when he pushed back his exuberant blond hair. Philip was bright. We had overlapped as students at Cornell. My wife was speaking to him animatedly. She placed her hand on his arm, and he looked over and waved at me. Philip then bounded quickly to the stage, his long arms swinging like a runner's. He held up his hand for the room to quiet down. Then he tapped twice on the microphone.

'Our speaker tonight has sometimes been described as the mad poet of particle physics. Starting with his first paper on quantum gravity, he has provided startling mathematical insights into the nature of our universe. He currently holds the chair in theoretical physics at Cornell University, a chair previously held by the great theorist Richard Feynman.

'Kobidi was born in Umudim, Nigeria, where he lived until he came over to the United States when he was twenty years old. He is the recipient of many awards, including the Dannie Heineman Prize for Mathematical Physics, and the Wolf Prize.'

Heidi walked towards me in an exaggerated slow-motion tiptoe, a file of papers in her arms.

'Professor Kobidi's work has often been criticised as arcane and dense, over-reliant on esoteric mathematics that fails to correspond to anything in the "real" world. In response, he once memorably commented: "Mathematics is the language in which everything fundamental is written. It is the poetry of the universe. Telling stories to explain theorems and the universe they describe is a movement away from the truth; ignoring the spirit and worshipping its graven image."

'Fortunately, Ezeani, at the urging of his beautiful and talented

59

wife, Heidi Kobidi, has started this lecture series in which he shares something of his ideas on the nature of our beautiful universe with a wider public. Please prepare yourself for a feast of ideas. Join me in welcoming to the stage Professor Ezeani Kobidi.'

The guests started clapping as he finished speaking and they turned to look at me, sitting on a raised bench beside my wife. She patted me on my thigh and gave me a long, adoring look. One that a member of the audience would believe conveyed deep reverence but also an abiding love. The type mothers reserve for difficult but gifted children. It was not a look my mother had ever given me. I stood up and walked to the stage.

In the dim light that fell on the gallery, I struggled to make out a face or two that I could focus on. After a few awkward moments passed in this futile effort, I closed my eyes, poked at the microphone and listened to the feedback sounds caused by the interference pattern in the electric circuits. Then I opened my eyes and began to speak.

'When I was a child, I had visions. A spirit often accosted me and spoke to me. The spirit was as real to me as you appear to me now in this room. Perhaps I had a very active imagination, or perhaps an undiagnosed psychiatric illness, or perhaps I *did* speak to a spirit?'

At this point Heidi's note said I should pause and smile. I was to expect some chortling. I did as instructed, and the small laughs came. I continued. 'Whatever its cause, this early experience planted deep inside me a foreboding that something deeper, portentous and unsettling, lay beneath the beguiling conventional surface of things. The world is not what it appears to be. Please look around. Where are you? Who are you? What are you?' There was a rustling in the group.

'When I started studying physics, I was a teenager. I came

to physics from mathematics, and I was stunned by the match between the equations constructed in our heads and what we see in the world. The real world.' I paused.

'However, this inherent beauty and order collapsed when I came to study quantum theory. It was as if I had been told an awful, shameful secret. I will let you in on it.

'We physicists have lost touch with reality. How has this happened? Well, simply because "reality" was never there. And we can prove it. Physics has arrived at a theory that accurately predicts almost everything, but this same theory makes a mockery of reality.' There were murmurs in the audience. I could hear a woman start to cough.

'Surely you exaggerate, Kobidi, you say? I certainly do not. To give you a sense of what is going on, I will describe to you three universes and ask that you tell me which one is: 1) a theory believed by a significant percentage of the world's top physicists; 2) a legend in an ancient religion; and, 3) a universe contained in a science fiction novel.

'Universe A is a universe where nothing is real. There are no objective events or things in this universe. However, in this universe there are lookers – who don't themselves exist, except when they are looking at themselves. The only things that exist in this universe are the phenomena that spring into existence at the gaze of the lookers.

'Universe B is a universe in which everything which can possibly happen does happen, immediately creating separate universes for each possibility. In this universe, if I toss a coin, one universe will be created where the coin lands on heads and another where it lands on tails. A version of me – if we can call it that – will reside in each separate universe, and these universes will never interact with each other again. New universes that replicate everything in the universe, all of us in this room, the

Earth, our galaxy – in fact, all the galaxies. The creation of split universes occurs not only when we do things on Earth. If an atom zipping around the outer planets of a star in a galaxy millions of light years away encounters an electron that can pass it on either the right or left, then both things happen and new universes are created with versions of all of us in this room.'

The auditorium was quiet, the audience riveted in attention. I continued to speak.

'Universe C is a universe where everything is one thing. In this universe there might appear to be separations and categories of things, but they are all illusions. The illusion of separation is a sort of conversation the universe is having with itself.

'Can you tell me which universe belongs to physicists, ancient religions or novelists?' I asked, and then paused as some in the audience shouted out different answers. The auditorium was animated, and some people repeated their responses. I cupped my hands around my ears as if I were encouraging them to speak louder. The audience grew even more excited, and I looked over at my wife. She smiled at me.

'I am afraid I have misled you,' I continued. 'Each of these universes is believed by a large number of the world's top physicists to be an accurate description of our universe.' I could hear a loud derisive laugh from some members of the audience; others smiled tentatively as if they were watching a magician, and they were uncertain the trick was over.

'How have ostensibly intelligent physicists arrived at these positions? What have their experiments and theorems shown that has led to this apparent drivel?' I leaned back on my heels. The room quietened. 'Well, let me tell you. The experiments and the mathematics agree, to an extent that is almost irrefutable. Every time we see a drop of water, once we look away

62

it is replaced by a spirit-sea of probability – inchoate, non-specific, and elusive. This spirit-sea is enormous and is not made of anything like water; in fact it would be appropriate to describe it as the spiritual potential of a water drop.' I stopped, then looked over the lectern, tapping a hand against its side. 'This quantum spirit-sea isn't made of anything like the stuff of molecules. The spirit-sea is made entirely of probability waves – similar to the waves of the ocean or the waves of sound that carry my voice to you and are governed by the same rules. The entire universe we live in is like this – when unobserved it behaves like a spirit-sea of probability waves. Most physicists believe that these waves – the probability waves – only turn into things – the molecules that make this lectern or the chairs that you are sitting on – when they are looked at or interacted with. This is everything. From electrons to galaxies, from humans to the moon. The three ridiculous-sounding universes I described are all attempts by physicists to explain this strange world that is real when we look and a spirit-sea of probability when our eyes are averted.'

I then started an explanation of the basic experiments that underpin quantum theory and why the results led to the strange universes I had described.

'If you shoot light through a double slit in a piece of card-board, it will behave like a thing – a particle – when someone is looking, and a wave when no one is watching. The same thing happens with an atom. These experiments have been performed millions of times, in every way human ingenuity can conceive. The results are always the same: things disappear when no one is looking.'

Then I provided a summary of the work I was doing and the approaches that had led me down new paths. The lively voices and animated shouts that had earlier filled the room had glazed into a resentful silence. There seemed to be relief when I stopped talking.

'Before I go, I will tell you something my spirit said to me. It is something that is said often among the Igbos, the people I am descended from: "When one thing stands, another stands beside it." Perhaps my spirit friend exists because I see him. Perhaps, in a certain sense, I exist because he sees me. Perhaps I am here because you see me. And you are here because I see you.'

There was quiet. It seemed no one was sure if I was finished.

Philip bounded on to the stage and grabbed the microphone. 'Please, a round of applause for Professor Kobidi,' he said. Heidi stood up to clap. She gestured to the audience to stand up too.

Obiageli Goes to School

After my father's inauguration as a professor of linguistics we had a large feast to celebrate. The table had been set up on the verandah, past the wide-open sliding doors of our living room. On top of the tablecloth were large serving bowls filled with rice and stews, eba, and pounded yam.

My mother handed my father a plate with rice, a red stew and large pieces of chicken and beef. 'Please, Professor, eat,' she said, beaming.

'Dr Okonkwo, don't jealous me o. Your turn is coming soon,' my father said to Dr Okonkwo, who stood beside him. They both laughed. Then my father lifted a spoonful of rice and stew to his mouth.

I looked past my father, over to the far corner where Professor Addo was sitting on the verandah's dwarf wall. He had a plate of food on his lap. Dr Yemisi walked over to him. 'How are you, Professor?' she asked.

'I am well,' Professor Addo responded. 'Please sit down,' he added, patting a place beside him. Just as Dr Yemisi sat, my father walked past me towards them. I was surprised to see Dr Yemisi immediately spring from the wall, stand and hug my father, placing her arm around his neck.

'Congratulations!' she cried and kissed him on his cheek. 'So well deserved!'

And Professor Addo nodded, 'Indeed. Indeed.' Dr Yemisi was still holding on to my father's hand.

'Hello, Ezeani,' Professor Addo called to me. He had noticed me looking in their direction. 'Are you still enjoying mathematics?' he asked. I stood still and stared.

'Answer Professor Addo. Don't be rude.' It was my mother's voice behind me. She placed her hands on my shoulders. My father's hand dropped away from Dr Yemisi's. Professor Addo was smiling encouragingly at me. I nodded at him, turned and ran into the living room.

My sister and her friend Margaret were lying on the settee with orange cushions, their heads facing opposite directions. They looked up at me — my sister having to turn her head — then continued their conversation. I sat down in the armchair in which my father usually sat and smoked his pipe.

'I don't understand why you would want to become a boarder. It is such a pain. Not when you live on campus and can continue as a day student. *Personne n'aime ça*,' Margaret said.

'There are too many people in this house,' Obiageli replied. '*C'est une maison de fous.*'

Margaret laughed. 'What did your parents say?'

'*Ma mere*, nothing. *Mon pere*: Why?'

'To which you responded?' Margaret asked.

'*C'est une maison de fous.*'

Margaret started laughing again, her laughter rising and falling in ever-louder peals.

My sister barely smiled.

My mother walked past the sliding door, striding towards the kitchen. She turned her head to look at me and the girls. 'Margaret, have you eaten? Let me make a plate for you. Obiageli, come and help me.' My sister did not move. Her face was set in an impassive stare.

'Didn't you hear me?' our mother said as she reached the kitchen door. My sister didn't respond. She didn't move. Margaret stared open-eyed at her.

'I don't have time today,' my mother finally muttered and turned to enter the kitchen. Through the open door I saw Ambrose, his huge bulk folded over a counter, lifting a large spoon of rice to his mouth.

I stood up and walked through the front door.

One cannot understand life without understanding time, and one cannot understand time without understanding space. The two are helplessly linked. We move relentlessly in the space that surrounds us, just as relentlessly as we move along in the time that surrounds us. The measure of each of our lives is the distance between the places, and the time between the events, in which we have appeared.

I walked out of the front door towards the shelf of grey rocks. Nnamdi, Chijioke and Gbemi had a large red-headed lizard trapped in a crevice. They were throwing stones at it. A few hit. The lizard ran desperately from one side of the hollow to the other.

'You guys can't aim. See!' Nnamdi said, and threw a stone that hit the lizard in its torso. Nnamdi laughed. The lizard, stunned by the impact, stopped moving. From the lizard's lower eyelid a translucent film rose, covered its eyes, and then slipped down again. Nnamdi threw another stone. He missed and laughed.

The lizard turned to me and said in a gravelly voice: 'They are playing a game and laughing. Meanwhile, I am dying here!'

I turned and looked at Nnamdi, Chijioke and Gbemi. They had not heard the lizard. Gbemi threw a stone that hit its head. 'Good shot!' Nnamdi exclaimed and then let out a loud laugh. The lizard's left eye hung strangely.

'I am dying here,' it said in the same gravelly, unrushed, voice. In terror, I turned and ran back into the house.

5

THE TIME I STILL HAD FRIENDS

Another House on Campus

The Standard Model of Particle Physics holds that there are four fundamental forces in our universe – the strong force, the weak force, the electromagnetic force and the gravitational force. The Standard Model is wrong; perhaps it is more accurate to say it is incomplete. There are only two fundamental forces – love and strife; the forces that bind things together and that which tears things apart.

Soon after my father was promoted to professor, we moved from the bungalow at the end of a cul-de-sac to a yellow two-storey building with a driveway that came in and out through cuts in a bougainvillea hedge. Behind the house lay a large yard with two Orji trees whose large roots reached like giant fingers into the ground. The house had four bedrooms and an enclosed study that my father lined with books. I do not remember the day we left one house and entered the other. Neither do I remember the packing and preparations that must have been made in anticipation of this change in our residence. As I remember it, we lived in one house and then, magically, we lived in another.

In the first memory I have of the new house, I sit on the white counter that runs along the walls of the kitchen. Maria's

— the maid's — hands are full of black-eyed beans and she is bent over a white bowl washing translucent skin off the legumes. Her mouth is open, and she is laughing. The laughter sounds like her voice, soft and slightly hoarse, like someone speaking when she is out of breath. I am telling her the story of a gifted boy in my class who farts in creative ways. Her eyes glisten with tears of laughter. She lifts the back of her hand to her face and in a short motion wipes the tears off.

My mother was standing at the cooker and turning plantain in a frying pan of oil. The distinctive smell, the smell of comfort, filled the room. My mother was also laughing.

'Ezeani, please let us finish making dinner. If we keep listening to your stories no one will eat in this house today,' she said.

'But that is not even the worst part,' I added. 'When the teacher walked back into class, the fart started smelling again.' Maria laughed even harder and for a second bent, with her hands on her knees, as if to keep herself from falling. My mother laughed even louder. The peals carried from the kitchen to the rest of the large house, moving in waves, which rose and crested in the molecules of air around us.

As if beckoned by this sound of joy, my father's head peered through the kitchen door. Immediately the door creaked, the laughter stopped.

'Why are you laughing like people in a madhouse?' he said. 'What is so funny?'

My mother's voice rose out of the silence. 'Ezeani is telling funny stories from school.' Her tone was measured. She breathed out. Loud enough so I could hear.

'Ezeani,' my father said, 'leave the women to their work. Let's go.' He made a motion with his hand to usher me out of the kitchen.

At dinner I sat with my father and mother. No one else was

at the table. My father had a book in his hand. When he was not lifting a fork to his mouth, he would hold it up to his face. My mother lifted moi-moi to my plate. 'Eat, Ezeani,' she said, smiling at me weakly and gesturing with her fork. Then she just stared at the food in front of her. My father rose, sipped some water, then left.

My mother stood up. She rubbed my head and then walked up the stairs to the bedroom she shared with my father. Her footfalls echoed heavily off the wood steps.

When I had finished my dinner, I left the dining room and walked past the living room to my father's new enclosed study. The room was filled with yellow light falling in twin circles from overhead bulbs. Another orb, from a table lamp, illuminated the professor behind his desk. He had a pipe in his mouth and sucked at it in a distracted way, an open book in his hand. The wall behind him was framed by a shelf that rose all the way to the ceiling, filled mostly with paperbacks, the spines a spectrum of colours fixed in the yellow light. I picked up a book from my father's desk and sat on the armchair in front of him. It was the same one – with the orange cushions – from the living room of our former residence. My father barely noticed. I folded my feet beneath me and started to read.

When I looked up, I saw Anyanwu sitting with his legs straight before him in the far corner. He had the checkered red, white and black cloth wrapped around his waist, his bare legs resting on the red carpet. Anyanwu stared intently into the pages of a leather-bound volume of the *Encyclopedia Britannica*.

This was what Anyanwu did now. Lurked around in the corners of rooms, reading and listening to conversations. He barely spoke to me. There were times when I would go days without seeing him. These were the days when I played with my friend Maria and my mother and laughed. Sometimes these happy interludes would go on for so long that I would wonder if Anyanwu had left for good. Invariably, when I started wondering, Anyanwu would appear in a corner, hitch his checkered cloth, and nod towards me in acknowledgement.

Suddenly, my father looked up from the desk at me and said: 'Chief Ezeani, what are you reading?'

I looked at the title of the book and read out: 'Euclid's *Elements*.' My father nodded in approval. 'You are reading the right books,' he said. 'What they won't tell you is that Euclid stole all the mathematics the Greeks use to distinguish themselves from the Egyptians! Now the Europeans parade it like their invention. What they call Pythagoras's theorem was first used by Egyptian clerks to calculate the area of land! They named it after one of their people and not the inventors.' He gestured towards the book. 'This is the work of your ancestors.'

In the corner, Anyanwu stopped peering in the encyclopedia and raised his head. He was watching my father and me closely. His lips closed and pressed tight together as if a question was straining to push between them.

My father took a draw from his pipe and blew the smoke out of his mouth slowly. The sweet smell that suffused the room intensified. I drew it into my chest. Then my father pushed an arm of his glasses into the mouth of the wooden pipe and snuffed out the flame. 'I am giving this up,' he announced sorrowfully. 'The evidence is incontrovertible. It causes cancer.' He started to pat the sides of the pipe against the large wooden ashtray on his desk, emptying it of half-burnt tobacco and ash. He smiled, stood, walked up to me. I watched him approach. He placed his palm on my head and rubbed it. 'You need a haircut,' he said, and walked out of the study. Soon, the sound of a jazz piano came to my ears, carried on waves of air from the living room. It was followed by a woman's breathless singing, her rasping voice rising and falling just before the notes of the piano. She sounded sad, like Maria. I closed my eyes.

My father stopped smoking his pipes. The sweet smell of burning tobacco no longer wafted through our home. I would find the discarded pipes in cupboards and drawers. I put these pipes to my nose, but there was no longer any sweetness, just an acrid, dry smell; bitter, like the resentful ash under cooking pots.

Anyanwu walked over to me. 'Let me see what you are reading,' he demanded. I pulled the book closer to my chest and ignored him.

'What is inside?' Anyanwu asked. 'Let me see.' He stood over me for a few moments and I pulled the book even closer to my chest. My mind was on the woman who sang like she was crying.

'Let us go and drink Fanta,' Anyanwu said, pulling me up from the armchair by my right arm. I followed him out of my father's study, and we walked, past the dining room, to the pantry under the stairs where my mother had stored sacks full of rice and beans, and tins of corned beef and sardines. Anyanwu pried open the locked door and we crawled into the tight space. Anyanwu took out a lobe of kola nut and bit on it. As he chewed, he lifted a bottle of Fanta out of a wooden crate, then paused and placed the neck of the bottle to his mouth, using his teeth to pry off the cap. He handed the Fanta to me. I put the bottle to my lips and gulped. Anyanwu opened another and drank from it himself.

The Bush That Ruins Little Birds

One day – this was during the long holidays, when schools were closed and Ambrose returned to Umudim and my sister Obiageli came back from her school's boarding house – my father drove past the bougainvillea hedge into the shallow driveway in a brand-new yellow Datsun, blaring the horn.

Nnamdi dismounted from the couch on which he had been hanging inverted, his legs in the air and his head pressed in the soft cushions, and ran to the front door. I followed him. 'Go and call your mother,' our father yelled, stepping out of the driver's seat of the still-running car. 'Let her come and see her new car.' As I watched Nnamdi turn around to obey my father's instruction, my mother appeared in the frame of the front door. My sister Obiageli was standing behind her.

Nnamdi beamed, his lips wide with joy. 'Nne, Daddy bought you a car!' he shouted, gesturing like one concerned he may not be believed.

My mother was not smiling. 'Did he?' she said.

My father arranged Nnamdi and me on the hood of the car, Obiageli leaned against the door pillar on the passenger side and my mother stood behind the driver's door, and then he took a picture with a Polaroid camera. As the picture emerged from the machine, he pulled it away and waved it in the air, like something that needed to be cooled. Nnamdi and I gathered to see the emerging family portrait, the image solidifying like reluctant ghosts taking shape. 'How does it work?' I asked, amazed and perplexed.

'It's just a camera, dummy,' Nnamdi said. 'Don't you know anything?'

There were other Polaroids taken that day. Pictures with my mother alone beside the car; one with my sister, her eyes on the ground, my mother's arm over her shoulder, beside the car's door; one with the hood of the car open and a neighbour pouring out schnapps or gin in libation.

In the afternoon, a young priest from Umudim who was

73

studying at the university came by. The hood was opened again, and the priest sprayed holy water, dipping a special stick into a vessel and flicking it at the car as he said a prayer. My mother held on to his arm. She made him stay and eat. He would not drink the wine my father offered. 'Jesus turned water into wine,' my father teased, and the young priest laughed.

'That is true,' he conceded, but he would still not drink.

'What is your thesis on?' my father asked.

'The Hermeneutics of Igbo Christology,' the priest replied formally, and proudly.

The priest and my father talked for a long time in a loud, animated way. They both seemed happy. I went to look for my mother.

As I walked towards the kitchen, Anyanwu grabbed my arm and pulled me into the storeroom under the stairs. The room was dark and its sloped ceiling, at the angle of the staircase, pressed down on us. I was frightened. It was the first time Anyanwu had laid his hand forcefully on me. I wanted to yell. Anyanwu's eyes were filled with rage.

'Don't be afraid,' Anyanwu said. 'Don't be afraid, my son, I will never hurt you.'

He put his hand gently on my knees.

'I am not angry at you. It is the idiot in the white cassock that has annoyed me,' he continued. He reached into his bag and took out a large drinking horn, filling it with palm wine from a tan-coloured gourd. He started to drink. He did not offer me a horn. As he drank, his eyes seemed lost, unable to focus. A copy of Euclid's *Elements* lay on a crate of soft drinks beside him.

'Let me tell you a story,' Anyanwu said, suddenly appearing to recover himself. I stared at him. I did not speak.

'The story is true. It happened many, many years ago. In the

74

time before the father of your father and his father and the father before him were born. But it happened to your fore-fathers, and it happened in Umudim.' He stopped, as if he had remembered something.

'I have not offered you anything to drink. You can have palm wine or, if you want, Fanta.'

I shook my head to decline.

'Are you sure?' he asked, his head cocked inquisitively. I nodded my head. Anyanwu nodded slowly in return.

'In Umudim there was once an evil bush that sprang up on the side of the river between Ibeto and Amaduru. You will not know the place, but it is easy to show you.' Anyanwu started to place his hand on my shoulder, but I brushed it off.

'Anyway,' he continued, 'the people in Umudim did not know that it was an evil bush. At first, it seemed like any other bush. It was only when children started to notice the tiny dead birds gathered around the roots that people started wondering. They called it the Bush That Ruins Little Birds. And it grew, and as it grew, misfortune in Umudim grew. The trees reluctantly bore fruit, meagre and shrivelled things; the yams were like cocoyams. It was not just the plants; everything born was small and gasping for life. Even children that were born in those times were smaller than children that were born in other times. The wise men in Umudim went to the bush and saw that it was not just little birds that died at its roots. They found the carcasses of small rodents, some that had not yet even opened their eyes, insects crawling from their larvae. Dead bodies littered the roots of this evil Bush That Ruins Little Birds. The smell was revolting. Everyone was perplexed. But it was when a vigilant mother noticed her infant daughter was always attempting to crawl away from her hut that things became clear. She decided to let the child go so she could follow and observe what was calling her.

As she walked behind the baby, her co-wives trailing after her, she screamed in terror and lifted her child from the path when she realised the child was heading to the Bush That Ruins Little Birds. The baby, her hands and knees scraped and red from the dirt and cuts, cried in terror when her mother lifted her, like one who had been woken from a terrible dream.

'It was then that the men of Umudim came to me. All I told them was this: brave men do what is required of brave men. This was a time when men were men. When they did not need to be led by the hand like children. They pulled the bush out of the soil by its stubborn roots and set it on fire. The bonfire burnt for eight days.'

Anyanwu stopped and nodded. I got the impression that he expected a reaction from me. I could not make any sense of his story.

At that moment, there was a soft knock on the storeroom door. My sister Obiageli opened it gently. 'Don't hide in here, Ezeani,' she said. 'Come, let's go and climb the guava tree.' She reached her hand out to me. I ran out of the storeroom and hugged her. She held me tight. She smelt nice, like something that was new and blooming. 'If you feel like hiding, Ezeani,' she said, 'just come and look for me.'

We walked out of the house to the guava trees in the yard. We stayed outside, hanging on the branches. Obiageli told me stories that made me laugh, till the sun started to dip below the clouds and Maria the maid called me to take my evening bath.

That night, when I had been long asleep in the room I shared with my brother Nnamdi, I was awoken by a curdling scream. It was a female voice. The scream came again, piercing and terrified. I could hear now that it was my sister Obiageli's voice.

Then there was the sound of a series of slamming doors and then there was silence. I lay on my bed, too terrified to move. Then I heard my father's voice muffled in a fierce whisper. My brother Nnamdi was still asleep, gently snoring. I lay awake on my bed. Eventually, sleep carried me away.

Common Entrance

Nnamdi stood over me. 'Get up,' he whispered. I sat up in the bed and he pulled me to my feet. 'Go and take your bath. You will be late for the interview. Daddy will drive me to school.'

My brother had left our primary school and joined my sister at the International School Ibadan. Unlike her, he was a day student, driven to school every morning by Taiwo, the driver that had been assigned to my father at his elevation to full professor and Head of Department. Unlike them, I had somehow conceived the idea that I wanted to go away. I had stood for the Common Entrance Exam that could get me admitted to King's College, the secondary school that I had selected, in what seemed a far-away Lagos. King's College, a school set up by the British in colonial Nigeria, was then, unambiguously, the most prestigious secondary school in the country. How I had developed the idea that this was the school to which I would go was unclear to me, and certainly to my parents. It was the first thing that I recall ever wanting, yet I cannot remember how or from where this desire arose. I was certainly not a competitive child, seeking confirmation of my academic prowess in a prestigious school. Perhaps it was the expression of a peculiar desire that I would only experience once in my life – the desire to move away to an imagined paradise; the compulsion to seek the transformation that would be possible at another coordinate in space, where time could perhaps, if not begin again, shuffle itself, and me, into something new.

The Common Entrance Exam was highly competitive, and I was not aware that much hope was placed on my ability to pass. It was therefore a surprise when the envelope arrived informing me that I had been admitted to King's College. I just had to show up for an interview. 'You are sure to get in now,' my father said. 'The interview is perfunctory.'

I needed to hurry, bath and get dressed. I could not be late for the interview. We had to leave early for Lagos, before the roads got busy. It was still dark when I joined my mother in the back seat of the car. I was sleeping again before we left our street.

The violent sound of metal tapping against the Datsun's window awakened me. When I had fallen asleep it had been raining, and the sun had not yet emerged. Now, the rain had stopped, but the road was still wet, the asphalt glittering. The sun had slipped above the horizon, but its light was low, muffled by the clouds, so the day seemed grey. I noticed these things before I took in the man rapping a metal bar against the window glass. He had a deep scar on his cheek and in his other hand he was waving a large revolver. I looked across to my mother seated beside me in the back. It was the terror in her face that finally caused me to understand.

Taiwo, the driver, opened the door, his head bent and cowered. '*Ejo, ejo*,' he kept muttering like a mantra, 'please, don't kill me,' over and over again. My mother pulled me to her as she opened the door and drew us out of the car. The men were yelling. I saw then that there were four of them. They jumped into the empty car; the man with the scar drove. The tyres of the Datsun squealed, and the car moved away from us at an amazing speed. It was only after it had disappeared in the distance that I noticed Taiwo lying prostrate on the road, red blood on his cheeks. At first, I thought he was dead.

Then my mother recovered her voice. 'Taiwo, get up from there before a car runs you over!' she shouted. We shuffled to the side of the empty road. We were only there a moment when twin beams of light appeared at a curve in the road, moving towards us. My mother started waving for the vehicle to stop. Taiwo still crouched.

'Madam, what if it is them? Please, let's just hide.'

'They drove towards Lagos. This car is coming from the opposite direction. It's not them.'

The approaching vehicle slowed and then pulled off the opposite side of the road. It was a white van. The front passenger door opened and a policeman in a dark blue uniform stepped out. My mother walked across the road and spoke to him. The policeman waved his arms in an animated manner, pointing in the direction in which my mother's car had disappeared. She nodded.

The policeman waved over to Taiwo and me, beckoning us. Taiwo held my hand as we crossed the road. The back door of the van slid open, running on a track that creaked. A small yellow light sparked, illuminating the glistening faces of the half a dozen policemen inside.

'Armed robbers are doing operation here,' the lead policeman declared to the others. 'Madam, I beg sit for front.' He walked over to the passenger side of the van and opened the door for my mother. Then he joined Taiwo and me, squeezed between three large policemen in the van's first row. When the back door slid closed, it creaked, then the weak light went off. The road curved over hills, the head beams throwing a yellow light against the wet asphalt. I looked out the window for my mother's car. There were no other cars. The policemen didn't speak. The only noises in the van were the wet sounds the rotating tyres made on the road and the policemen's rhythmic breathing. Soon, I fell asleep again.

While I slept, I dreamt. In my dream the robber with the scarred face was gleefully driving, laughing, while at the same time he rifled through my mother's handbag. The teeth in his mouth were a vibrant yellow. Two upper incisors were missing. Suddenly, his companion, who had also been laughing, his feet on the dashboard, shouted a warning – but it was too late. They had driven off the road.

Now the four robbers were gathered around the blown tyre, squinting in the weak light. A blow from Anyanwu lifted the leader above the bonnet of the car and into a ditch. Anyanwu was wearing only a loincloth and his red, white and black checkered cap with the eagle feather. His muscles were tight, and his chest heaved as he drew breath. The robbers scattered in alarm, running off to the hills that looked down on the road.

The dream ended then. Taiwo was gently nudging me awake. I was confused. The large policeman beside me was smiling, his yellow teeth flashing between his lips. Taiwo pulled me out of the van. It was only then that I saw the white walls, a blue gate and the white colonial buildings. We were outside the gate of King's College.

Visiting Day

There are two fundamental forces – love and strife; the force which binds things together, and that which tears things apart. This epiphany would come to me not as a solution to the problems that would later bewilder me as I pushed past the early success of my work in quantum gravity. It would come to me as a simple insight, illuminated by the friendship I once had with a boy. A breakthrough that opened my eyes to the importance of entanglement to our understanding of our universe.

Nothing means anything except in connection with something else. We live in a universe that consists of nothing but relationships. A thing can only be if it affects something else. It is this relationship between the thing that stands and the thing that stands beside it – this is what being is. On my first day at King's College, I walked through the gate and was escorted by a senior student past classrooms, the dining block and then, passing the cricket oval, we climbed up to the second storey of the assembly hall to my assigned dormitory, Panes' House. The bunk beds in the hall were arranged in rows set against each wall and between them was a row of lockers. The sight of the white buildings, the careful lawn of the cricket oval, and the senior student in a white uniform and a blue blazer filled me with a sweet excitement. I had arrived at a new place. And with a naïve notion, I assumed that new meant good.

A voice called out, 'All freshers, come here right away!' and I was to learn, barely ten minutes in this new school, that I was mistaken. I travelled towards the voice till I saw its source, a tall, light-skinned boy. 'What is my name?!' he bellowed. The nine or ten freshmen boys stared at him. No one spoke. It seemed no one knew his name. 'What is my name?!' he shouted again. 'All of you kneel down, till someone tells me my name.'

We all went down on our knees on the concrete floor. It was the voice of a slight boy who seemed to swim in his white uniform that saved us. 'Begha, please,' he said.

The boy who knew the prefect's name was Ope Adesola. Once we had been released by the Prefect Begha, we sat in the dorm quietly chatting among ourselves. Ope surprised me and said: 'I know you from Ibadan. I was in your class in Class Three and my brother Gbemi is your brother Nnamdi's friend.' I could not remember him from the University Staff School, but he would become my best friend at King's College.

On exeat day, Ope and I walked towards the gate, a line of students in front of us. We were all wearing our school uniforms: white shorts and white shirts with the crest of the school on a badge pinned to our chests. My mother waved from the yellow Datsun parked across the road. Obiageli was leaning against it. Her hair had grown out into an afro.

'Ope, how are you?' my mother said. 'Get in, get in, we are all going to enjoy.' Obiageli got into the passenger seat beside Taiwo the driver. Ope sat between my mother and me in the back of the car.

'Did you meet any armed robbers on the way?' I asked Taiwo, smiling.

'God forbid!' Taiwo exclaimed.

'It's not a joking matter,' my mother chided. 'I was so lucky that the thieves had a flat tyre and just abandoned the car where the police could find it.'

My mother took us to a restaurant that served grilled fish and sticks of fried yam on large plates. We feasted on the food and Fanta. 'You guys are like a pack of wolves,' Obiageli said, smiling.

'Let them eat,' my mother retorted.

Just before 6 p.m. the Datsun pulled up at the school. Obiageli got out of the car and hugged me. My mother waved, her eyes moist. Ope and I strolled languidly to the blue gate.

Pains House

The library hall at King's College and its rows of books arranged on wooden shelves became my refuge. I would come into the room and sit at a table. The sun threw light through the windows at an angle that changed as the day moved from morning to evening. In the library, I was forgotten. The librarian, a large woman who seemed to dislike books, would leave frequently to sit in the administration block and gossip. Occasionally, a group of boys might stroll through the library door, their voices rising and jarring, but soon I would hear the door open again and their voices falling away as they left. I found solace in that library, among the dusty shelves and old periodicals, sitting, reading through books.

One afternoon I had just settled myself at my favourite table in a corner, unobservable from the front door or the main reading room, when I heard a sound behind me. I turned and saw Anyanwu. 'The Greeks were not comfortable with infinity,' he announced. Then he threw Euclid's *Elements* on the table. It was the same edition as the one in my father's study. On the cover, a bearded man with a large cloth thrown over his shoulder drew with a stick on the ground while a young student crouched attentively. I ignored Anyanwu. But I knew what he meant. I had started working through some mathematics on my own.

The door opened. 'Is there anyone here?!' It was Begha's voice. He started walking down the rows. I hid under the reading table. Anyanwu had disappeared.

'There is no one here,' Begha said. 'Come! Kneel down.' The person he had been speaking to emerged from the book stacks. 'Sorry, Begha, please,' he said as he knelt. I knew from the voice before I saw his face. It was Ope. 'I really fetched the water, please. Sorry, Begha, please.'

'Adesola, open your mouth,' Begha said, struggling with the zip of his pants. He released his penis from his white trousers with his right hand. Then he slapped Ope hard across his face. It was then that Ope opened his mouth.

The dorm was filled with bustle. It would soon be lights out and bedtime. I looked over to Ope at the end of the dorm. His face was fixed and vacant. As I walked over to him, he turned away from me towards another boy and started to tell a joke. They both laughed. Their laughter was loud, so loud that some of the boys in the dorm stared. Close to lights out, loudness could attract reprimand or punishment. Begha ignored them. I turned and started walking back to my bunk.

'Kobidi, where are you going!?' Begha shouted at me. I stopped and looked at him. 'Where is the water I asked you to fetch for me, Kobidi?'

I had no recollection of having been asked to fetch water. But I could not be certain. 'Begha, please,' I said. 'I did not fetch water for you.'

'Kobidi, you are a lout. You are admitting it without contrition.' I looked at Begha, then I looked down at the floor.

'Kneel down,' he said.

I raised my face and stared at him. I could sense that everyone in the dorm was now looking in our direction. Nothing seemed to happen for a long time.

'You are still standing!' Begha shouted. 'In fact, go and mount, Kobidi. Right now!'

'Yes, Begha, please,' I said as I clambered onto one of the lockers and stood as still as I could. I closed my eyes. Soon the dorm grew quiet. I waited for Begha's instruction releasing me. It did not come.

Most of the dorm was soon asleep, the sound of snoring trilling in the dark. Sleep washed over me, snatching parts of consciousness. The mosquitoes bit at me and sometimes the sting would bring me back to wakefulness. But only momentarily. I fell deeper into sleep. The stretches between spans of consciousness grew longer. Then I could see a wide desert of black sand and a turbaned man with a yellow cape waving disdainfully at me. He seemed to be shooing me off, like I was a stray dog.

My tongue tasted the saltiness of the floor and the iron of blood. A sharp pain struck at my jaw and shoulders. It took me a moment to realise that I had fallen from the locker onto the floor. My jaw and arm were in excruciating pain. No one else was awake. No one heard my fall.

I tried to lift myself. Words fell out of me, and I did not know what they meant. I was blind with tears and rage. I stretched my arms to find my bunk. As I lifted myself onto the upper bunk, I woke the Form Two student in the bed below. Startled, he muttered: 'Stop shouting, Kobidi. It's the middle of the night. Anyanwu is not in Panes' House. I think he is in Hyde-Johnson.' It was then I realised I had been calling for Anyanwu.

I stopped shouting and fell into bed. I had barely fallen asleep again when I heard Begha's voice whispering in terror: 'Don't kill me. Please don't kill me.' It was difficult to see what was going on in the darkness. My eyes began to adjust. In the weak light Anyanwu held Begha's hands together so he couldn't move them. The look on Anyanwu's face was hard and his jaw muscles twitched. Anyanwu beckoned to Ope as he slowly placed a knife to Begha's throat. Ope sat up in bed, his eyes wide open. He stared hard at Begha. Anyanwu shifted the knife near Begha's throat, teasingly. Ope turned his back to them and then began to cry. Anyanwu shrugged, took the knife away from Begha's throat and remorselessly sank it into his shoulder. Begha screamed. Anyanwu clapped his large palm over Begha's mouth as he drew the knife out. 'Don't kill me!' Begha whispered, looking into Anyanwu's face. Some of the other students had started to stir. Anyanwu slipped away. Begha was now screaming at the top of his voice in terror, kicking his feet in the air madly.

Anyanwu walked over to me. He moved his face close to mine. 'You called me, Ezeani. I, Anyanwu, have answered you,' he said. Then he placed his palm on my forehead. His hand was calloused and hard, but on my forehead it felt cool. The commotion in the dormitory grew louder. The senior boys got the housemaster. Begha was taken to a hospital. Then forgetfulness, gratitude and sleep washed over me like they were concurrent waves, the moments between them fixed.

In the morning Begha was back in the dormitory. I was told by some of my classmates that the doctors said if Begha's stab wound had been a few inches to the right he would have died. Others said the wound was superficial, like one a small child might make with a penknife, and it was only Begha's panic and the sight of blood that had caused the trip to the hospital. There was a great fuss in the school and whispers of investigations. There was no final determination as to who had stabbed Begha, but the motivation of several junior boys, whom it was clear he had assaulted, was deduced. Begha's father, a high court judge in a city hundreds of miles away, was required to make an appearance in the school for a conference with the principal. In the end, Begha was removed as a prefect and banished from the boarding house. He would become a day student, compelled to live with relatives of his father.

Very little was said of that night. I would form the impression then that, for many of my schoolmates, the most important thing was to avoid the entire truth being known. It is an impression I have since had in many other places. It is an idea I have never understood; the idea that the most important thing is not to know the truth.

6

FLUID DYNAMICS

Entanglement

When particles interact, their quantum wave functions briefly overlap. In quantum mechanics, once this happens those particles are forever linked: a single wave function describes both particles simultaneously, a process known as quantum entanglement.

Ope sat with his feet on my bed. He was wearing his uniform, white shirt and white trousers and polished brown sandals. His eyes were focused on the wall. He threw a red ball, watched it smack against the wall and rebound to him. He caught the ball and, hardly waiting, threw it again. He caught the rebound.

'I don't know. I think I'll go to Ibadan and study history.' He threw. 'That's exactly what my father did,' he added, as he cocked his head and caught the ball. He threw the ball and the sound it made as it hit the wall echoed through the empty dormitory. 'What will you study? Engineering?'

'I will study mathematics at Ibadan,' I said.

Ope missed the ball. He got up to pick it off the floor and looked over at me as he put his hand around the red sphere.

'Why?' he asked.

'No surprises. I know mathematics and I know Ibadan,' I said.

A junior boy wandered into the dorm, someone in the second form.

'What are you doing here?' Ope asked, his voice tightening with authority.

'Prefect Tagbo asked me to tell you that the other prefects are waiting for you in the Prefects' Room, Adesola please.'

'Oh yes,' Ope said, gathering himself. 'Go straight to your class. Don't loiter.' He waved the student off. 'Kobidi, I trust I can leave you unattended,' he said with a small smile and walked out of the dorm.

After the initial success of the papers I wrote while I was working on my PhD, the only other great advance in my work as a theoretical physicist would be the insight that fell into my head when I began to understand the true meaning of entanglement. It was an understanding that would come to me whole and complete. An epiphany. Like all truly profound truths, we have all stumbled upon it. You must have sensed it too – that we are connected to the world and each other in ways that transcend time and space; something deep links us.

When it came to my mind, I was sitting bare-chested on the deck of my home in Ithaca, New York, near the campus of Cornell University, reminiscing on the day when a friend I once had, Ope Adesola, was beckoned by Anyanwu to slit the throat of another boy and Ope turned away and curled his back.

It was not immediately clear to me what my epiphany meant. I did not understand that the work which followed – work that utilised entanglement to expand my initial equations on quantum gravity – would lead many of my colleagues to question my science, and ultimately my sanity. Derision and calumny would be thrown at me, my reputation eviscerated, and spurious errors attributed to the most elegant equations I would ever write. Equations so beautiful I would often read them and weep.

At the height of these trials, my wife would choose to inform me of her intention to file for divorce.

Although many of my colleagues will disagree, I am simply describing the way our universe works. You, I know, have sensed it too. An uncomfortable reminder that the world which we struggle to believe in is fundamentally a fraud. But perhaps *fraud* is too strong a word, perhaps it is more precise to say that it conceals far more than it reveals. No less clear-seeing a physicist than Einstein was tortured by this. Perhaps you know that he referred to this entanglement, a phenomenon that fell from his own mathematics, in disbelief as 'spooky action at a distance'. Yes, until his death, despite the physics he could understand better than most, he would insist that 'God does not play dice'.

I, Ezeani Kobidi, because I had an opportunity to live with a Sun God named Anyanwu, knew that this was not true.

The Sum of Things

When multiple forces are applied to a body, their vectors can be added, and if the sum is zero, there is no net force. Consequently, there is no movement.

I returned to Ibadan, to the campus where I had grown up, and started my study in the Department of Mathematics, Faculty of Science. A building that was approximately a ten-minute walk from my father's office and a twenty-minute walk from the second house on campus, where my parents still lived. I would go over often to my parents' house on weekends and Maria, who, strictly speaking, was no longer the maid, would make me lunch, while my mother sat on a couch in the living room looking out of the window.

'I have made your favourite soup, Ezeani,' Maria said. 'Yam pepper soup with dried fish.' She smiled and gestured for me to sit at the dining room table where a single setting was placed. I grabbed the spoon beside the large bowl, cutting off pieces of yellow yam in the brown broth and carrying the laden spoon to my mouth. Maria sat in a chair beside me, a slight fragrance of roses and musty earth clinging to her body.

'How is your lawyer?' I asked.

'He is OK. He is paying me,' she said.

She watched me eating. Then she smiled and asked: 'Ezeani, nnam, how are you?'

I looked at Maria. My mother was still in the living room. In the world that I lived in before boarding school, it was my mother who would sit next to me, her hand on mine, while she smiled.

'Nne,' I called to my mother, 'are you charging Maria rent? Her lawyer is paying her a lot of money.' And I laughed loudly, so my mother would know that it was a joke. But my mother did not speak. There was no way to know that she had heard what I said.

Maria looked at my face and whispered: 'Come and see your mother more. She barely talks to anyone now. Sometimes, out of nowhere, she will start abusing me!'

Shortly after I left for King's College, Maria, released from chores related to my siblings and me, had completed secretarial college. She could now type sixty words a minute and knew Pitman shorthand. Arrangements had been made by my father for her to work as a secretary for a lawyer who had an office on the top floor of a three-storey building in Bodija. She still lived with my parents. Every weekday she would leave for work early in the morning and return at 3 p.m. to do chores around the house. Maria had taken over the duty of making dinner from my mother. When Ambrose had finally completed his Masters – three years later than scheduled – he had left Ibadan for a job my father had arranged for him at a teachers' training college not far from Umudim. Maria had moved into the vacant boys' quarters of the two-storey house immediately Ambrose left. She had bought a flower-patterned fabric and made a heavy curtain for the window and door of her room.

I lifted some more of the warm food to my mouth. When I looked up, my mother was standing at the entrance to the dining room.

'Maria, you are becoming lazy with your work. I don't know if it is the secretarial school we sent you to that is deceiving you. Please get up and go and prepare the beans in the kitchen.' Maria rose from the dining table and walked through the kitchen door without saying a word. The rose scent left with her.

My mother sat in the vacated chair. Her skin was taut and the hollows below her eyes were dried out like a broken gourd. I held her hand. Her fragrance was gone. Now she had the vague, smoky smell of dried fish. The transition had been slow. I had barely noticed it while I was away at King's College, only observing that her visits slowed down and then stopped. When I returned to Ibadan the backs of her heels were dried and cracked.

'My son, Ezeani, Lord of Men, how are you?' she said.

I smiled. 'I am well, my mother. How are you? You don't look like you are laughing enough.'

'I am laughing, my son, at the things that are funny and the things that are not.'

I smiled.

'I worry about you, Ezeani,' she said. 'Is your stomach still bothering you?'

'Sometimes. It comes and goes,' I replied. 'The doctor at UCH gave me some medicine and it helps.'

'Let us hope it works. Watch where you eat, though. You never know what is in a stranger's food.'

I looked at my mother. I was overcome with a sudden, deep sadness. But I could not tell why. My stomach growled.

'Nne, I am going,' I said, and leaned over to hug my mother. 'I am going back to campus.'

My mother looked surprised. She held on to my hand. 'Let me give you some food to take with you?' she said.

My mother held my hand as we walked up the stairs, turning right at the landing to go into the bedroom she shared with my father. The curtains were drawn. A large fan spun slowly over the bed on a giant hook fastened to the dark wooden beam that spanned the roof. The air molecules barely stirred. The room was stuffy, filled with motionless scents. My mother reached into her blouse and pulled a necklace of keys over her head.

'We have thieves. So, we lock things,' she said. She shuffled to a large black metal case and opened two padlocks. Reaching in she took out biscuits, cocoa powder and tins of condensed milk.

I set the polythene bag of provisions my mother had given me onto the dark floor of the university dorm room and then threw myself on my bed. My head hurt. It had happened before. The headache would strengthen until I was incapable of moving. My stomach tightened and I could feel gas filling it up, visibly distending my clothes. As the pressure built, the pain increased, until gas loudly passed from my anus. The sweet joy of release from that awful force was mixed with embarrassed, wet, round noises.

I lay in the dark and groaned. My mind skipped from one image of my mother to another. She smiled in each image. Sometimes her face was dry and tight. My cheek was wet with tears. I did not understand why I was crying. My stomach bit at me, so sharp that I pulled my knees towards my chest and groaned. Then I heard Ope opening the door. He stood in the lit frame. 'Are you OK?' he asked, before he walked into the room.

Ope had arranged for us to be roommates in Mellanby Hall. Our small room had two single beds forced into opposing walls and a small study desk placed against the window that looked onto the Quadrangle. Ope pulled the chair from under the desk and sat down. 'Are you sick? Do you want me to take you to the clinic?' His voice was deep, assured. Even here in university, he kept the air of easy authority that had contributed to his selection as Prefect of Panes' House.

I shook my head. Ope stared at me for a few minutes. 'Look, I am going to have to leave soon,' he said slowly. 'Just came back to take a shower. I'm riding my bike over to Chijioke's house. She's having a party. Her mother travelled, so it's really going to be big.'

After his shower, Ope returned from the bathroom with a white towel wrapped around his waist. He stopped at my bed and put a palm to my head. 'You don't have a fever,' he said. 'At least it's not malaria.' I closed my eyes. I could hear him dressing, then the hiss and fragrance as he sprayed cologne. As Ope walked out of the room, I let out a loud, large fart. He continued through the door, pretending he hadn't noticed, his cologne's thick musk clinging to the air behind him.

*

When multiple forces are applied to a body, their vectors can be added, and if the sum is zero, there is no net force and so there is no movement. However, if the body is a fluid, complexity increases dramatically. Analysis is no longer trivial, and the sums are almost never zero. A fluid is a state of matter that yields to sideways or shearing forces. Liquids and gases are both fluids. The equations of motion for a solid body are a small set of simple algebraic expressions. The equations of motion for a fluid are a large, complex and intricate set of partial differential equations.

Alone in the room, I started farting almost continuously. It seemed that with Ope gone my efforts at suppression had come to an immediate end and the built-up pressure was determined to find release. Instantly, the room was filled with a rich, loud laugh. I farted again. I could not control it. The laugh became louder. The peals moved towards me in increasing intensity until Anyanwu's amused face emerged from the darkness in the light reflected from the Quadrangle.

His appearance had not changed. He pushed back the goatskin bag slung over his shoulder and continued to laugh until the laughter became a short cough. I turned away from him and looked at the wall.

'You shouldn't laugh at sick people,' I said.

'You are not sick,' Anyanwu said.

'I have IBS,' I said. 'Irritable Bowel Syndrome.'

'There is nothing wrong with you. Let us drink.' He reached into his bag and brought out a large gourd of palm wine stopped with a mash of green leaves. As he lifted the stopper, the frothy, rich smell of the palm wine filled the room. He poured out a full horn and handed it to me. I took it from him, but I did not drink. He filled his horn and emptied it in one movement and then he wiped his mouth with the back of his hand and belched.

I laughed. 'Drink!' Anyanwu commanded, and started to fill his horn again. I put the horn to my lips and drank.

When we had drunk many horns, Anyanwu spread himself on the floor beside my bed, looking up at the ceiling, his eyes open, lost. As I settled my head, I felt Anyanwu's coarse hand grab my arm. 'Let us go out!' he declared. I looked at him. 'Don't worry, I will dress up,' he said, and lifted himself from the floor.

The cinema was in an open field in Bodija, behind a road that passed by the railroad track, miles beyond the gates of the university. There were about a dozen plastic chairs arranged in three rows before a makeshift white tarpaulin screen. Behind the chairs, a projector, placed on a tall three-legged stool, cast a beam of speckled light. A tall, gaunt woman stood beside the projector, selling alcohol from large plastic gallons. She would pour out measures into plastic cups, sometimes mixing herbs in customised brews. I had never been to this part of Bodija before. The dings of stray rocks echoed against the metal siding of empty warehouses. A crowd of men jostled for space and alcohol. Coarse language and rough voices filled the air, rising often to threaten violence.

The film played to shouts of derision and applause. It was a Chinese film that depicted a courageous band of friends fighting off kung-fu vampires. Often, women's clothes were torn off in battle, revealing bare breasts. A cry would go up from the audience. 'These Chinese papaws small o!' one man yelled over the sound. There was loud laughter. Anyanwu laughed loudest, slapping his thighs through the yellow corduroy trousers he was wearing. His eyes shone and his teeth caught the light from the screen. On his head was a yellow fedora, a large, white eagle feather stuck in its band. He shook his head with mirth.

On the screen, the lead vampire had been trapped by the kung-fu heroes and was steadily losing the fight to their complex moves. Suddenly, and for the first time in the movie, the lead vampire turned into a bat, escaping imminent defeat. Anyanwu immediately stopped laughing. His face was filled with disgust. He rose. From behind, several men shouted at him to sit down. Purposefully he walked to the screen, held by strings and clamps, put up both his arms and tore it down. There were screams of anger and disbelief. Several men rushed at Anyanwu with bottles. They surrounded him, shouting at him in Yoruba and Pidgin. Anyanwu spoke back in Igbo. This seemed to infuriate the men even more. One of them lunged at him. Anyanwu expertly dodged the blow then struck the man hard across his face. Then he turned to me and shouted: 'Run!'

I ran.

Anyanwu and I lay hidden in a brush at the side of the railroad track, watching the men as they wandered in the darkness searching for us. 'Let me catch that idiot!' one of them threatened. The others muttered vile words. One of them struck a bottle against the ground. We held hands and laughed silently, our eyes shining like little balls coated with tears. I pulled hard on Anyanwu's arm. He looked at me. Then he put a finger to his lips, signalling me to stay quiet. My stomach was calm, the pressure dissipated, as if someone had twisted open the valve on a gas canister.

I woke up beside the railroad track. It was very early; the sun was still barely visible, but the horizon was alive with refracted light. A train conductor in his uniform kicked at my leg. 'Young man, get up. This is disgraceful. Get up before anyone else sees you.' There was vomit on my shirt. Anyanwu was gone.

I stood up and walked downhill, away from the railroad track, towards the university.

Obiageli Slips Away

'How are you settling in?' my father asked. He was seated at the desk in his office. Large piles of books lay on his desk and the floor. 'It is important to make sure your first year goes well. It sets the tone for everything else.' I stared at him. He smiled, perplexed.

'Sit down,' my father said. 'You look tired and unkempt.' I sat down on the chair in front of his desk. 'Do you want something to drink? A coke?' I barely nodded my head. He leaned to his side and called loudly to his secretary to bring in a Coca-Cola and some biscuits. 'And some tea,' he added.

I was still drinking the coke when my father asked: 'Have you heard from your sister?'

I nodded. He waited for me to swallow. 'She calls me every Sunday at the Porter's Lodge at the Hall and she writes to me often.'

Obiageli was at the University of Maiduguri in the north. It was the university as far from Ibadan as it was possible to be while remaining in Nigeria. Still, it was not far enough. Obiageli did not return home on vacations. I had just received a letter from her telling me she had received a scholarship from the University of Connecticut and would be leaving Nigeria to take it up.

'What is she saying?' my father asked, his voice rising in pitch in a way that made it sound like this was a casual, offhand inquiry.

'Nothing really. Just the usual,' I lied.

'I wish your sister would be more open,' my father said. 'She is my only daughter. I worry about her. I don't know why she is so intent on separating herself from the family.' My father's eyes seemed wet, as if he was holding back pain. I also felt pain and I realised that it was sorrow transferred from my father.

I thought: I feel sorry for him. I wondered what I might say to my sister so she would be kinder to him. I wondered if my father was aware of my sister's plans to leave Nigeria; I wondered if he just wanted to find out if I knew.

'You are the only one that she seems to care about,' my father said, looking at my eyes. I shifted them. 'Her relationship with your mother is certainly awful. And there, I am sorry to say, I believe it is your mother that's at fault.'

I nodded. Perhaps my mother was at fault, I thought. I didn't know. Maybe she could be kinder to Obiageli.

'Your mother has been a huge disappointment,' my father continued, shaking his head.

I was uncertain how to respond. I picked up another biscuit and ate it slowly. My father looked inquisitively at me. He lifted the teapot and poured tea into a white porcelain teacup. Then he leaned back in his chair and started telling me the things that were wrong with my mother.

Later that evening, when I was summoned to the Porter's Lodge to take a call, I was surprised to hear Nnamdi's voice on the line. He was calling from the Student Union at his university in Nsukka.

'Did you hear your crazy sister is going to America?' he yelled. 'I have to say, you must admire it. You just have to admire it. I wouldn't mind going to America myself.'

'Why would you want to go to America?' I asked.

'That's where the action is.'

'There is nothing you can do there that you can't do here,' I said.

'You are naïve my boy, quite naïve,' he replied. 'Speaking of your naivety, have you had sex yet? You have to do it in your first year or you might doom your entire university career. My advice is to find the first girl that will say "yes" and get it over with.'

I did not respond.

'OK. I am going to take that as a "no". When I come back for the long holidays, if you haven't managed to do the deed, I will steer you in the right direction,' he said, and then started laughing. I was quiet.

'Don't be so serious, Ezeani. I need you to help me with something. I need Nne to send me some money. When I call the phone at home, no one picks up. Can you go to the house and tell her?'

'Why don't you call Daddy in the office and tell him?'

'Because I don't want to ask him,' he replied curtly. 'Please old buddy, ol' pal, do your loving elder brother a favour. Will you?'

'Yes,' I responded.

'Great! I have to go. I will call tomorrow to find out what Nne said.'

Before I could respond, I heard the click of the phone line disconnecting.

When I returned to my room the lights were all on. Ope was seated like a Buddha on his bed. Chijioke stood at the sill looking out at the Quadrangle. The light that passed through the window lit the left side of her face. She turned to the door as I stepped in.

'Well, well. Young Kobidi,' she said 'You have grown up. How are you? How is your brother Nnamdi?'

'He is fine. I just spoke to him,' I replied.

'And how are you?' she asked, smiling at me. 'Ope says you couldn't come to the party because you were sick. Hope you are better?'

'Yes,' I replied.

Chijioke lifted a tape cassette from the desk. 'I believe this is what I was promised,' she said, placing the tape into a back pocket of her jeans. 'See you later, Ope. Say hi to your brother. See you later, Ezeani. Say hi to your brother.' Then she gaily walked through the door.

'Where did you go last night?' Ope said, immediately the door shut behind her. 'When I got back from the party you weren't here. I was really worried.'

I threw myself on my bed and grunted.

'The headache is back,' I declared.

'All I am saying: next time, leave a note, Kobidi.'

Nnamdi Seeks His Pay

I barely lifted myself from my bed that week, my pillow covering my head. Messages came with increasing frequency from the Porter's Lodge, informing me that my brother was on the line. Sometimes I let out a non-committal grunt, other times I did not respond at all. Most of the time I stared at the wall, but sometimes I read a few pages of Euclid's *Elements*. I could smell my body turn and start to give off a rancid, cloying odour. Ope came in and out. He would look at me sometimes and ask: 'Do you need anything?'

Towards the later part of the week, Ope pulled the pillow from my head. 'Look, Kobidi, I'm not sure what is wrong, but your brother is really bugging me. He keeps calling me at the Porter's Lodge. He wants to know if you've given some message to your mother.'

'Can you get me a coke and meat pie?' I responded. Ope nodded his head and left the room.

On Saturday morning I awoke at dawn. The sun's rays were only starting to come through the window, but I was filled with a strange, vibrant energy. I leapt from my bed, threw a towel over my shoulder, and walked down the hall to take my first bath in a week. As I rubbed soap on my belly and lathered my body I felt like singing. Instead, I declared: 'Let's go forth into the world, dear friend. Let us go forth.'

Ope was still sleeping when I left. I started to walk towards my parents' house, taking the shortcut between the sports fields. As I walked past the sculpture garden at the Institute of African Studies, I saw a handful of birds picking at something in the grass. I smiled and stopped to watch them, sitting myself on a low stone fence across the street. A red-tailed bird landed on the ground; it looked at the earth and then picked at it. Abruptly, the bird trilled and flew into the air. A larger black bird with a white collar of feathers around its neck landed on the head of an abstract figure of a Benin king, squawking and fluttering its wings.

As I watched the birds, I suddenly felt the brightness and joy that had filled me when I woke diminish. I immediately had the thought that it was too early to go over to my parents'. I rose and turned in the opposite direction. I had no clear destination, but as I walked, I noticed that I seemed to be making a large loop that would eventually take me back to my parents' home.

The university campus was so familiar that each street, or trail through a field, threw off memories that I could taste in the back of my throat. The succession of these memories, the taste of them, seemed to improve my mood and I could sense the joy to which I had woken returning to me.

I took a long, winding path past Awba Dam. The campus was quiet, and I walked in the grey dawn in this solitary way, sometimes smiling to myself at the nostalgia that swam up with the passing streets. When I was walking past the university bookshop, the sun's rays intensified, and the grey dawn lifted. I looked across the avenue to the Seat of Wisdom, the Catholic chapel where my mother had taken us when Ambrose first came to live with us. The doors of the church were open. I crossed the road.

The church was empty. The wooden pews faced a large stone altar. Behind the altar rose a green and yellow wall of stained glass. I knelt in the last pew, close to the door. I could sense that I wanted to ask for something, some relief, but I was not sure what I wanted or in what way I should ask for it. I bent my head. My ears were instantly filled with the screech of birds. Then there was silence.

I heard a rustling sound and lifted my head. Before I could identify the source of the sound, I heard my name. 'Ezeani, is that you?' Then I saw the priest from Umudim who had blessed my mother's yellow Datsun. He was thicker now, walking to me briskly, his legs shuffling his cassock as he moved.

'My son, why are you here so early?' he asked. 'Are you OK?'

I noticed on his face that he had stopped because he could now see there were tears in my eyes. I did not know when, but I had started crying.

The priest moved closer to me. 'My son, it will all be well in Jesus's name,' he said. 'Let us pray together.' As he was placing his arm around my shoulders, a loud blow landed on my ear. My face seared with pain and I became momentarily disoriented. I pushed the priest. I could hear his feet shuffling away from the pew.

'Leave him alone!' a loud voice bellowed. The priest stood erect; his eyes fixed on something behind me. I turned. Anyanwu was wearing only his loincloth and the floppy white, red and black cap with a large straight eagle's feather affixed on one side.

Anyanwu grabbed my right ear. The pain was piercing, and I turned my head to lessen it. The priest, who had retreated halfway down the aisle, started approaching again. In his right hand he held an aspergillum, and he was throwing holy water at me and Anyanwu. The priest began to mutter some words in Latin. There was a grim, resolute look on his face, yet he seemed frightened.

'If you don't stop that nonsense, I will make you drink that water through your nose!' Anyanwu shouted. He spoke to the priest in the distinct Igbo of Umudim.

The priest paused. 'Ekwensu! Ekwensu!' the priest shouted.

Anyanwu laughed. A short, mirthless laugh. 'How dare you call me by the name of that weakling?' he snorted. 'You are a fool. Yes, you, son of Amadike, grandson of Ikedimba, great-grandson of Ihite. And you come from a long line of fools.'

As Anyanwu recited the priest's lineage, the priest stopped moving altogether.

'How do you know these things from Umudim?' the priest asked.

'Who are you questioning? If I open my eyes and you are still here, I will feed you that stick you are holding in your hand.'

The priest moved back a few steps but did not leave.

I was dragged out of the church by my ear. The priest did not attempt to follow.

Anyanwu pulled me across the road and threw me down on the lawn. 'What is wrong with you? How can you bend your knee to another spirit?' Anyanwu yelled. 'Why are you behaving like this?'

Anyanwu was shaking me. I was getting angry, and I swung at him.

He let the blows land.

'When it is me, when it is Anyanwu, the Spirit Lord of Light, that's when you have the courage to fight.'

He laughed a mirthless laugh as I continued to hit him.

I noticed that a small crowd of students had gathered and were watching as I flailed.

I didn't care.

Anyanwu had started talking again. I could not focus on what he was saying. He was angry and his Igbo words fell in such a torrent that I had difficulty gathering their meaning.

'We do not pray by begging on our knees,' he said, grasping my shoulders. 'We do not let the spark of our Chi, our divine light, be dimmed by obsequious conduct. You are old enough; you should understand the span of your spirit!'

I noticed that the crowd around me was growing and some of the students had their hands over their mouths.

'Look at me!' Anyanwu commanded. 'How can you learn when no one teaches you? Listen to me, so you can understand!' I swung at him again. He let the blow land on his chest, then he blew a froth of saliva and herbs into my face.

Suddenly, I was enveloped in darkness. There was no sound. I reached out my arms, but I could not see them. Then the feeling that I had arms went away. I felt quiet and then alone.

Next, as if it was a slow warmth, I could feel Anyanwu beside me even though I could not see him.

Anyanwu spoke: 'I will teach you what your ancestors knew. Calm your mind, Ezeani. Calm your mind and listen.'

'What nonsense are you talking?' I tried to speak. But there was no mouth, no lips, just the cloying gloom.

'Calm your mind,' Anyanwu continued in a soothing voice, as if he had heard me. 'Listen, when the world was still upright, these were things taught to little children; yet you, who has started growing hair under his chin, does not know that the universe is made of Chi and Eke. Everything in the universe is of these essences, whether they are living or dead, whether they are alive or inanimate. Perhaps you have heard of the Chi, the life force that provides everything with its own reason for being?' As Anyanwu's voice vibrated the darkness, a single yellow light emerged, flickering and then burning a steady flame. Soon there was another light and then another, and then thousands of yellow lights sparked in the void as far as my eyes could see.

As the lights jumped, I could sense my eyes start to take form behind the pulsating landscape. Gradually, the rest of my body returned to me, the drumbeat of flames went out and I was covered in darkness. In the congealed black, I could hear Anyanwu's voice.

'The universe unravels itself in phases. In the phase before the present one, the universe was one thing, a singularity, called Oma. Then the Chi of Oma, Chi Ukwu, separated itself from the Eke of Oma and we entered the current phase of the universe in which everything is in twos. The Chi and the Eke. The present universe is ruled by the Parent God Chi ne Eke, Chi and Eke. The two go together, in this universe. Chi is the animating spark or force and Eke is the manifestation of that force in existence. Everything in the realm of Chi is potential and spirit, and everything in the realm of Eke is manifestation and existence.'

My mind was clouded, and I tried to hold on to the words that were falling from Anyanwu's mouth. But his voice had started drifting away. I could hear my name echoing, 'Ezeani, Ezeani,' from beyond the void. The sound kept repeating itself. Soon the voice was the same as my mother's voice. I opened my eyes and saw my mother's face. My head was in her lap. Around us in a semicircle were a dozen university students, whispering.

'Reverend Father called us,' my mother explained. 'They said you had a spiritual attack.' I gazed at her. Her eyes looked so sad that I looked away. She soothed my head with the palm of her hand. We stayed there for a long while and then a few students helped her lift me into the yellow Datsun. My mother drove me home.

Mother Leads the Way

The man in the brown uniform took off his hat and pressed a dark purple towel to his brow. Obiageli gazed at him. '*Oya*, bring it,' the man said, barely looking at her. She reached into a brown manila envelope and brought out white forms and a few passport photographs.

'You are here for international passport?' the man asked.

'Yes, the only type,' Obiageli said wearily.

'What was that?'

'Yes, international passport,' Obiageli said.

I looked across the waiting room. There were perhaps a dozen people. We had come early and walked in at 9 a.m. when the office opened. It was getting hotter now, and looking at the sun through the open window in the hall, I concluded it was just after noon.

'Ah! You fine o!' the man in the brown uniform exclaimed. He held Obiageli's passport photograph up in the light.

Obiageli looked away from him, pushing her eyes to the floor.

As we walked towards the yellow Datsun afterwards, Obiageli smiled. 'The passport is done! Next – visa – and I am on my way!' She slid a flat metal key into the driver's door. Like an eager plant, a metal plug rose from an encircled metal hole in the door panel. I jumped into the passenger seat. Obiageli carefully backed the car out of the parking space. A young man in a red Manchester United jersey was waving and giving directions. When we had pulled out, he tried to get a tip, pushing an empty hand towards my sister. Obiageli ignored him. 'Touts,' she said as we drove off.

'What is studying mathematics like? I mean, is it intellectually gruelling? Soul-sapping?' Obiageli stopped talking and looked at me. We were approaching the university gate. Obiageli drove cautiously. A few students waved at us to stop. Obiageli kept going; she was waiting for me.

'You know you could study something else?' she said. 'I mean, no pressure.'

I looked at my sister and I smiled. 'It's actually the easiest subject for me,' I said.

'Perhaps,' Obiageli said. 'But you shouldn't put yourself under pressure. I hope you don't buy that nonsense from your father about academic excellence.'

'I have no idea what you are talking about,' I said.

'I'm just worried. I heard about the IBS and that you aren't going to your lectures. Ope says you are sleeping a lot.'

I didn't speak. I looked at the dashboard. The plastic was brittle and old, with a brown patina that came from the dust that had settled deeply into the polymers' fibres. My brother Nnamdi had replaced the original radio with a shiny unit with equalizer displays that pulsated in purple. I played with the radio dial.

'Do you know about Chi?' I asked.

'The Igbo Chi? Like in Chi'm na luru mu ogu?' Obiageli replied.

I nodded.

'Well, it's like your personal divine spark, your spiritual counterpart. I don't know. Something like that,' she said.

'Do you have a good Chi?' I asked.

Obiageli laughed. 'My Chi is always fighting for me! Not like your lazy-ass Chi that sleeps all day.'

I looked at Obiageli and smiled, then I looked down at the floor mat.

When I looked up, my sister was staring through the windscreen. She took her right hand off the gearbox lever and placed it over my hand.

'Don't worry, Ezeani. You will be OK. We will all be OK.'

And then she smiled.

'When will you leave?' I asked.

'In about a month,' she said. 'Perhaps you can join me when I am settled in America. I am sure you can get a scholarship too.'

I scoffed.

'Of course you can,' she repeated.

I smiled. I really did not know whether I could get a scholarship or not. It was a prospect that belonged to a set of possibilities that I simply had never considered after King's College. The possibility of leaving Ibadan, the possibility of willing myself into new things.

Obiageli swung the yellow Datsun into the driveway of our house. She was laughing at a joke I had told, her eyes gleaming and her mouth open. Immediately we passed the bougainvillea, the smile disappeared. She slammed on the brakes. The car came to a quick stop. Obiageli leapt out. I jumped out too, reaching the front door before she did. The policeman at the door grabbed Obiageli as I slipped through. Another policeman was running after me. 'Stop! You can't go up there!'

I was already at the top of the stairs. My father and a large policeman were standing on the landing. They looked shocked. I ran past them. Behind me, my father called, 'Ezeani!'

I was already inside my parents' room. My mother hung by a thick polypropylene rope on the hook that held the ceiling fan. I walked to her. I looked up at her gaunt face. Her face looked neutral, as if she were indifferent to a choice that lay before her. I touched her feet. They felt hard. Then I felt the policeman's arms around my chest. Then there was another policeman. They pulled and dragged me. I felt I was yelling but I don't recall anything that I said. They were pushing and pulling me down the stairs.

My father was at the bottom of the stairs. He came to me. He hugged me, folding me into his arms. 'Be brave. Be stout-hearted,' my father said. I was surprised. He was speaking to me in Igbo. I laid my head on his chest and cried, the tears falling from my eyes unrestrained. My father placed his hand on my head. He patted me. A policeman was still holding my sister's arm. She was shrieking. My father started walking towards her. As my father approached, she shook her arm free. She raised her right arm and pointed at my father.

'You bastard! You fucking bastard!' she yelled. My father stopped. 'Get the hell away from me,' Obiageli said, and turned to walk out. At the door, she paused. A policeman stood close by, as if he was unsure whether to block her path. My sister looked him in the eye, and he stepped back. My sister then turned and looked at me. 'Ezeani! Come, let's go. I am not going to spend another night under the same roof as this murderous devil!'

'Obiageli, calm down,' my father said. 'Calm down. Your mother's depression has been getting worse. No one expected her to do this. Calm down.'

'Get the fuck away from me!' Obiageli screamed. The coarseness of my sister's words shocked me.

She stood in the doorway and beckoned to me. 'Ezeani, come. Let's get out of this madhouse!'

I sank to my knees on the dining room floor. Then I bent over and crumpled to the ground. My father rushed to my side. He knelt and hugged me. 'Ezeani, be stout-hearted,' he said again in Igbo and patted my head. My sister stood in the frame of the front door. She waited there for a few seconds. Her arms dropped. My nose was filled with my father's scent. Then my sister turned and walked out.

I lay on the floor, and I cried. I would never see my mother again. I did not know it then, but it would be over a decade and a half before I would again see my sister Obiageli.

PART II

PART II

THE MIGRANT BIRD

Migration Patterns

It was winter. The sound of branches creaking under the weight of accumulated snow, and then cracking with a sharp snap, came to my ear in successive waves and for a moment dominated my senses.

I opened my eyes when there was silence. My sight was filled with whiteness; large snowdrifts between the trees; fluffy white lines balanced on branches; and new snowfall, specks of white drifting to the earth. I heard the cooing of a bird and lifted my eyes to scan the bare branches. I couldn't see the bird. And I couldn't see the sun, hidden by a sea of sterile grey clouds. The bird cooed again.

I closed my eyes and reflected on sound – these waves, produced when our mouths, cooing birds, a breaking branch, pressure molecules of air. Sound is an emergent property of pressure waves. What we hear is not the fundamental thing: even though they may deceive us, voices are unreal; just a construct that emerges in our mind from pressure dissipating in air.

The small group that had been walking behind me, speaking in hushed voices, the binoculars and cameras hung around their necks nodding against puffy bright-coloured winter coats, had turned back towards the Visitors' Centre. As a cold wind started, I passed the wooden sign, covered with uneven snow, which read: *Cornell Lab of Ornithology – Sapsucker Woods Bird Sanctuary*. I pulled the ends of my large blue coat together and zipped it till it hid my sweater, then looked down at the tops of my black boots speckled with fallen snowflakes.

As I drew the frigid air through my nostrils, I tried to organise the thoughts leaping in my mind. I pushed my cold hands into the pockets of my coat. My right index finger passed through a hole in the fabric and I could feel the cool layered polytetrafluoroethylene of the coat's exterior. In my head, I arranged, as if around the perimeter of a box, the equations that would be at the heart of my PhD thesis: 'The Fundamental Equation of Quantum Gravity'. I had decided on this title earlier in the morning, looking in the mirror with a toothbrush in my mouth.

'Rather ambitious,' Professor Lantier whispered, a wry smile on his face, as he took a sheaf of papers from me. The office was overheated. I unzipped my blue overcoat as he gestured at the pale wooden chair in front of his desk. The radiator in the corner clanged. I sat down. '*The* fundamental equation?' he asked, his smile disappearing.

'Yes,' I responded.

Behind his head was a framed London tourism poster with a picture of Buckingham Palace. The caption read: **When a man is tired of London, he is tired of life.** Lantier looked up from the papers and fixed his eyes on me. Then he smiled again, a wide smile. The smile pushed the corners of his moustache and beard further apart. I shifted in the chair before his desk. 'From what I understand, your project is to utilise concepts in basic mathematics, concepts in combinatorics, especially permutation, to simplify Feynman's complex algebra into a simple one-line equation?'

'Yes.'

'Combinatorics mathematics is really basic stuff. Feynman's equations call for many pages of quite complex algebraic calculations to describe even the most basic particle interactions. I am not aware of anyone who has suggested that basic concepts such as permutation can be utilised in physics,' he said. In the silence I looked down at the floor. Professor Lantier turned to the sheaf of papers. He was reading intently. He turned a page and placed his right index finger on his lower lip.

'If I understand this correctly,' he said, gesturing at the papers, 'you expect to arrive at the solution by extension of this simple method?'

'Yes,' I responded.

He seemed lost in thought for several minutes, his face raised to the ceiling and both his hands grabbing the arms of his chair. 'That would eliminate time from the equations. That's what the simplification implies. You can still arrive at the right answers without factoring *when* things happen,' he said suddenly, his face moving down till he was staring at me.

'Yes,' I said, even though I was unsure he had asked a question.

'Give me a few weeks to review this,' he said. 'It is certainly a very fascinating idea.' I stood up, turned and began to walk out of his office. 'This is amazing,' he continued. 'You are suggesting that time is an emergent property! If you can demonstrate this, you are on your way to becoming one of the greats.'

I continued past the frame of his door. There was nothing I did that would cause him to believe I had heard him.

I made my way out of Clark Hall towards Stewart Avenue, where I knew I could catch the city bus to the Sapsucker Woods

Bird Sanctuary. I sat on the bench, under an awning built to shelter waiting passengers, and closed my eyes. There was a direct shuttle that ran every hour, but I knew I had missed it. I dwelt on the missed shuttle and a feeling rose in me like a mix of diluted heartbreak and regret. Then for a moment I wondered why I had thought of this missed shuttle at all. It was useless information. It could have no effect on what I did next, but the bitterness of regret lingered in my mind. I heard a couple approach the awning. They stood, speaking, beside the bench. Her voice was loud, with a lilting tremble that seemed like hope. My eyes remained closed.

A loud hissing and clanking caused me to open my eyes. It was the sound of the bus's brakes and the pneumatic pipes and cylinders which drove them. The white bus bucked, its front rising, falling and rising again, as it pulled to a stop. Then there was another hiss and the bus doors opened. The couple climbed on. I got up and followed them. The steps were covered by a perforated black plastic mat framed by silver metal. There was dirty snow in the corners. The driver looked at me as I climbed aboard. His broad, round, pink-tinged face seemed to shift as if in disapproval. 'Fare?' he asked loudly. I reached into my pocket, took out a plastic card and pressed it against the plexi-glass screen that enclosed him. He barely glanced at the card before he turned back towards the windscreen. The bus started to move. Its forward motion jolted me, as if I had been pushed from behind. I regained my balance and sat in the empty seat behind the driver. I looked around. There were nine people in the large bus. The couple sat at the back, holding hands. I dropped my head.

One cannot understand life without understanding time; and one cannot understand time without understanding motion. The motion of each body through space can only be measured in time.

The motion of each body through time can only be measured in space. And one cannot begin to understand motion until one realises that there is no absolute motion. Nothing moves except in relation to something else.

I closed my eyes and floated in the spaces within myself. I felt like I was in a cloud, its fluffy borders absorbing time, keeping me in an unending, uncertain present. Then the bus started to brake, forcing me forward in my seat. I opened my eyes. We were approaching a bus stop.

The man behind the desk at the Visitors' Centre nodded at me. I knew from experience that he would be behind the desk on Tuesday, Thursday and Friday. He wore a white button-down shirt with a brass nametag pinned to his chest. The first few buttons on his shirt were always undone so that his black chest hair and a gold-coloured crucifix were visible. The name tag said *Jesus Mendoza*.

'Back again, Doc!' He had a loud, melodious voice that sat on the vowels. I nodded, then showed him my pass.

'Aw, you don't have to do that, Doc,' he said, as he waved me through. 'You are one of our regulars.' I pushed past the turnstiles and walked out into the cold air.

The cooing song came again to my ears and I scanned the canopy of leafless branches. I couldn't see the mourning doves. I knew, from reading a pamphlet at the Visitors' Centre, that it was unusual for these birds to remain in Ithaca this late in the year. They were one of those species of migratory birds that moved between the southern and northern hemispheres seeking warmth, sustenance and a good place to breed.

Then there were two birds cooing. I could feel the cold on my face. I knew if I continued thinking about the birds I would start crying. It had happened many times before. I would see something which at first seemed innocuous, and then a deep sadness would envelop me, like thick smoke wrapping itself around my chest. Once, sitting on the bus, I had watched a woman strapping her child into a stroller. She was wearing a thick coat and her back was turned to me. The child was giggling, and the dark smoke filled my lungs. Was it foreboding? A sense of the possibilities, the potential ways the mother would betray this child? Or the inverse: the potential ways the child would betray the mother? The way this moment would be betrayed by other events we would think of as the future? I do not know. I did not know when the tears came. I bawled, my eyes rapidly dripping tears, and my nose filled with snot. 'Are you OK, sir?' a teenage boy holding up a skateboard in the seat across the aisle asked. I had got off the bus at the next stop and started walking in the cold to my room at Telluride House. Telluride was the student residence I would live in through all my years at Cornell, all but the years after I met and married Heidi. Upstairs in my room, I sat at my desk, placed my bare feet on the wood floor, pulled out a handful of blank sheets and started work on what would be my first paper: 'Relativity, Symmetry and the Consistency of the Universe'.

I closed my eyes again, pushing out the thought of the cooing birds. I drew a box in my mind, setting the equations on the box's edges, so that they would crawl from one end of the box to the other and then make a turn and continue, till they came back to the place they had started. I played with the equalisations in my mind, moving the symbols to a balance. I moved them automatically, along unconscious paths that bound the connections of one thing to the other. Suddenly, something new appeared to me. It took me a moment to notice its meaning. A relationship that I had not seen before. It was an epiphany. I stopped walking and stood still in the middle of the path. I cannot tell you how long I stood there. I could see clearly that as the equation crawled to the borders of the box and turned, these were the corners where time turned to space. 'One has to be careful not to say *when* time turns to space,' I said to myself aloud. The sound of my voice surprised me. It was the longest sentence I had uttered in two years. I bit my lip, pleased no one had heard me speak. Draughts of cold air hit my face and whipped the drifting snow into flurries. My ears were beginning to grow cold. I started walking again.

The Children of a Lesser School

Before my PhD and my initial success in physics; before the ruin of my reputation close to the moment when I entertained my most profound insights; before I began to discern the meaning of time and consequently the meaning of my life, I went through the tumult of undergraduate study at Cornell. Yes, there is no other word to describe that time – especially the last year, when I was arrested by officers of the Ithaca Police Department.

As I emerged from these events and returned to Cornell to start my PhD, I would build bulwarks against the terror and confusion that had threatened, again, to destroy me. The first of these was a purge of silence. The only words that anyone would ever hear from my lips during the three years that followed would be 'yes', 'no' and 'undetermined'. The words represented the limited scope of the information that I wished to convey. They also represented the exercise of a right that I would learn during this process belonged to me – the right to remain silent – guaranteed to me by the constitution of the United States.

In those undergraduate years, Anyanwu and I appeared to lead separate lives; we rarely appeared in the same place at the same time, our lives only linked by the slightest thread. Anyanwu sought to assert his influence on me by feigning indifference. I would see him occasionally in Collegetown, dressed in varied costumes – in the garb of a UPS delivery man, in long shorts throwing dice on a concrete outdoor court with a posse gathered around him, and once, at the back of a restaurant, in a tweed jacket, speaking animatedly to three sneering women. When I approached him, he would feign bewilderment, turn his eyes away from me or walk off rapidly, looking back nervously to ensure I was not following. He seemed to have made a lot of friends of his own and barely sought my company.

Anyanwu was most persistent in one particular disguise. I would see him often on the benches at the edges of the square in Collegetown, always with a small coterie around him, a gold-coloured Yankees baseball cap on his head. He seemed a bit taller, but a flash of recognition would glint from his eyes as I walked past. One day, tiring of this absurd pantomime, I approached him directly. As I raised my arm and greeted him, he did not turn or walk away, but laughed at me and then shouted loudly: 'Dude, you are crazy! I don't know you! Get the hell out of here before you get hurt.'

He hadn't fooled me. Even though he spoke in English, with an amateurish approximation of an American accent, I heard the Igbo in his voice. Occasionally, I would see him at a club or bar. I noted that he seemed to keep himself away from the facilities of Cornell University and I only saw him in Collegetown or once at Ithaca College, a less prestigious university, on the other side of Collegetown. At Cornell, Ithaca College was generally viewed, derogatively, and in all likelihood unfairly, as a 'party school'. This designation suggested its students were not focused on academic excellence but on drinking, socialising and sex.

Anyanwu's aloofness towards me only changed during my last

year of undergraduate study. This followed the afternoon, late in the summer – a period when the campus was largely empty with most of the students still away on holiday – when officers of the Ithaca Police Department came to interview me regarding my knowledge of Anyanwu's activities. Then, he sought me out. It was then that Anyanwu appeared at the foot of my bed, shadowed me through my classes, sat at my side while I ate my lunch, and earnestly and angrily implored me to keep my mouth shut and ignore the requests and summons of the police.

In those undergraduate years, despite the tumult Anyanwu introduced into my life, his attempts to dictate, distract and dispute, his insistence, petulance and mockery, I learnt physics. I learnt the nature of the world we live in. And unlike many of my course mates, I actually believed what I was being taught.

You might wonder what I mean when I say this. It is this simple: we live in a strange universe that deludes those who think of it as a place represented by what we see, what we hear and what we touch. I listened to my professors, I read the texts, and what they described was frankly, if one is sincere, fantastic. The common path, among those who studied and learnt this physics, was to understand these things, comprehend this dismantling of the common-sense notions that hold our ideas of life together, but to then continue living as if ignorant. I, on the other hand, believed what physics said. I believed because the proof, in mathematics and experience, was clear. I did not make the mistake of many of those around me. I did not take these things as a manner of metaphor, a mathematical description of complex systems that did not alter the fundamental nature of my daily life. I listened and I believed. And I acted as one who believes must act.

Breakfast (with a Smile) in Collegetown

The music was low. Over the loudspeakers, a voice came on to announce that the station played the best of today's and yesterday's easy rock. I toyed with the salt shaker on the table in the booth. The diner was all stark colours. The walls and floor were white and the benches in the booths red. The tabletops were all a bleak, black, hard Formica. The waitresses smiled often, twirling the skirts of their uniforms of checkered red and white aprons. And they were beautiful. A knife and fork sat on a paper napkin in front of me. None of the waitresses had come over to the booth I had selected in the back.

I pushed the salt and pepper shakers together. They collided, transferring energy, momentum and velocity in the way the laws of motion predicted. I pulled them apart again and pushed them towards each other with even greater exertion. I could sense within myself a spirit propelling me with increasing violence. It was seeking to impart a destructive force, one that would shatter the lattice cohesion of the glass shaker's molecules and spread salt and pepper and glass shards all over the Formica tabletop. As the salt and pepper containers moved at an apparent fixed rate towards each other, a bell rang at the front of the diner. I looked up. Nnamdi walked through the front door in a camel-coloured overcoat. He was wearing a black turtleneck and blue jeans. I noticed he had on dark brown boots. His eyes immediately locked on mine, and he smiled. He looked elegant. Like a picture one might see in a magazine. The salt shaker missed the pepper and flew off the tabletop, shattering on the floor.

Nnamdi walked confidently towards me. Without breaking his stride, he motioned to one of the waitresses, pointing at the

mess of salt on the ground beside me. She nodded in acknowledgement.

He was a very handsome man. It was here, in this diner, that
I saw this clearly for the first time. I realised, as if discovered
with the borrowed vision of the blonde waitress, that he had
occupied himself with exercise which had built up his chest and
arms so he looked like a man who had been cut for something
higher.

He seemed to be waiting for me to do something. Was he
expecting me to stand? To hug him? He pulled off his coat and
hung it on one of the hooks beside the booth. I hadn't noticed
the hooks. I hadn't taken off my blue polytetrafluoroethylene
coat. Nnamdi nudged me. 'Move over, Ezeani,' he said.

I slid further into the booth. Nnamdi sat beside me. A waitress appeared. She leaned over a long-handled broom, sweeping
the salt and glass shards into a dustpan. Then she looked up at
Nnamdi and asked, 'What would you guys like to drink?'

'A black coffee for me and an orange juice for my brother,'
he replied. 'Sorry for the mess.' He smiled at the waitress. She
pushed her hair behind an ear and smiled back at him, her eyes
widening with her grin.

Nnamdi turned his attention to me. 'How are you, buddy?'
he asked. I stared at him.

'Are you OK?' he asked.

'Yes,' I replied.

'Traffic was hell getting out of New York. The roads in winter
can be tricky. It was a good drive, though. Scenic. It's good to
get away from the city,' he said.

I didn't respond.

'Ithaca is nice. I'll be here for the whole weekend. We booked
a room at the Statler.'

Nnamdi looked around the diner. His gaze seemed to linger on one of the waitresses. Then he turned to me.

'I came up with someone,' he said. 'She's waiting for a car to move so she can park the BMW. I was afraid you'd leave if we were late. You should get a phone. They're really useful.'

The bell rang again. A tall woman with light brown skin walked through the door. She was beautiful. She didn't have an overcoat. She also had on a black turtleneck and jeans, but tucked into black, knee-high boots. Her hair was piled high on her head and strands of it fell in ringlets to her ears. Her earrings reached her shoulders. My brother raised his arm to beckon to her.

'She's from Trinidad,' Nnamdi said, as if he had explained everything.

She slid into the booth across from me. 'Hi, Ezeani!' she said and smiled broadly. Her fragrance filled the air with the scent of flowers crushed in light oil. 'Your brother has told me so much about you! He says you are brilliant.'

'Almost as brilliant as me,' Nnamdi added, a laugh on his lips.

'Yeah, of course,' the woman said, rolling her eyes at me. 'My name is Dawn.' I shook the offered hand and smiled at her. She was so pleasant.

'I have heard so much about you. I've been dating your brother for two months and you are the thing he is most proud of. After you, it's his BMW.' She smiled, as if we were sharing a secret.

The waitress placed the orange juice in front of me with a straw. Then she put the coffee in front of my brother with two small containers of milk and a teaspoon. 'Do you guys know what you'd like to order?' she asked, looking at my brother.

'I'd like a coffee,' Dawn said. The waitress turned to her, barely nodded, then returned her gaze to my brother and smiled. Nnamdi smiled back, then he looked away, and faced me.

'Have you looked at the menu?' he asked.

'Yes.'

'I think I will have an omelette and toast. Should I order for you?'

'Yes,' I said, pointing with a finger.

'And what will you be having, my lady?' Nnamdi asked Dawn, his eyes fixed on the menu.

'Pancakes and maple syrup,' Dawn said, her voice like a song.

Nnamdi repeated the entire order to the waitress, including Dawn's request for coffee. The waitress wrote on a small pad and said, 'Coming up,' cheerfully to my brother. Dawn smiled a small smile, then turned to me.

'Do you like Cornell?' she asked.

'Yes,' I responded.

'Are you happy here?'

'Yes.'

'Do you think you will stay at Cornell after your PhD?' she asked.

'Undetermined,' I said.

Dawn had a puzzled look on her face.

'Since he started his PhD, Ezeani only says "yes", "no" or "undetermined". Don't worry, you'll get used to it,' Nnamdi said. Then he turned to look at me. 'Of course he's staying at Cornell. They would be fools to let him go! I've been reading up on this. My brother's papers are getting cited. He's the hottest PhD candidate on the East Coast. Maybe there is one other guy at Caltech in the same league.'

'It's incredible how accomplished you guys are,' Dawn said. 'Mr Wall Street over here, you, the hottest physicist, and your sister, the paediatrician. Your father must be really proud.'

'Mr Wall Street?' Nnamdi scoffed. 'You make me sound like a stockbroker. I trade complex derivatives. Brain work.'

'Excuse me,' Dawn laughed. 'Touchy, touchy.'

I laughed. It was so unexpected that I looked around the room, surprised.

'Did you hear about Daddy and Maria?' Nnamdi asked.

'No.'

'They are getting married. Can you imagine?'

I didn't respond.

'Your father is remarrying?' Dawn asked. 'Who is the bride?'

'Our former maid,' Nnamdi said. 'Father has been fucking the help.'

'What?!' Dawn said, a quizzical look on her face. She was looking at me as if she wanted me to confirm my brother wasn't misleading her.

'Yes,' I said.

Then she reached her arm across the table and held my brother's hand.

'Can you believe Daddy asked me what I think?' Nnamdi laughed. 'Like that makes any difference. I don't want to even imagine what Obiageli will say!'

'Your sister doesn't like Maria?' Dawn asked.

'My sister has daddy issues,' Nnamdi quipped, in a careless, throwaway tone. 'Anyway, I told him he should go ahead. It's his life. Have some fun while you are it.' Nnamdi paused and looked at the tabletop.

'Aren't you hot, Ezeani? Do you want to take your coat off?' Dawn asked. When she opened her mouth to speak, I noticed the redness of her lipstick, and that it had stained her two front teeth.

'No,' I said firmly, my voice hostile and perceptibly louder.

Dawn seemed startled. Heat travels from warm to cold bodies, never the other way. Heat is the only distinction between the future and the past; between what is living and what has died. She looked at Nnamdi and smiled a small, embarrassed smile. Nnamdi squeezed her hand.

Then he let go and reached across me to pick up the menu propped lengthwise against the wall. 'I want waffles too,' he said. He beckoned to the waitress.

Bodies at Rest

The door of my room was closed. I could hear my father's voice on the other side. But I could not hear what he was saying, the wood absorbing the vibrations of the air before it could reach my ears, changing the amplitude of what remained, so it came to me as a muffled noise. My father pushed the door open gently and his head appeared in the narrow crack. 'Will you come to the office with me today? We can stop at the staff club and you can speak with Professor Addo. I gave him some of the things you've been scribbling and he is really fascinated.' His face looked so hopeful.

I wanted to say yes, but I turned around and faced the wall. 'I don't want to go out,' I said.

My father stood at the door for a few seconds more. Then he walked into the room. I turned around and looked at him. His face was taut, his eyes wet, filled with pain as he looked around. He started to pick up the clothes that littered the floor. He folded some and hung others in the wooden wardrobe. I turned away and stared at the wall. I could hear the clinking sound as he picked up some of the dirty dishes on my desk, crusted with days-old rice and cornflakes. As he opened the door, I turned and saw his arms laden with three plates and three dirty glasses. On the desk, there were still four dishes with half-eaten food, caked with red oil. My father closed the door gently, a low sigh forcing itself from within him and trembling the air in my room.

Behind the closing door, Anyanwu glowered. 'Get up!' he howled at me. 'Look with your own eyes! Don't be afraid. Write it all down. You are surrounded by liars. Don't listen to them. Listen to me.' He was wearing the same white, black and red checkered cloth, but instead of the usual cap, he wore the fedora. I turned away from Anyanwu. He walked to the desk, picked up one of the dirty dishes and threw it with great force to the floor, shattering it into a kaleidoscope of shards. As I stared at him in amazement, he fixed his eyes on me, picked up each of the remaining dishes and in turn smashed them on the floor. A few minutes later, my father returned. He looked at me and I turned away. He left the room again, but this time he did not close the door. Then, after what seemed a very long time, he returned with a broom and a purple plastic dustpan. In silence, he swept up the broken dishes.

These were the years before I left Nigeria and arrived in the United States, when I lived alone with my father and Maria in the second house on campus. I would sit at my desk and scribble equations and poetry on sheaves of paper, non-stop for hours, and then, abruptly, roll into a ball and sleep for days. I barely left my room. I never went to class. My degree was on hold and it was Professor Addo and my father who ensured I was not simply expelled from the Department of Mathematics. My siblings, Nnamdi and Obiageli, had moved to the United States. My father would come into my room and lay out letters from my sister. The letters were always thin and flat with those unimaginative stamps that were essentially cropped pictures of the American flag. I read all of Obiageli's letters, but I cannot now tell you what she wrote. Only that in each letter there would be a pamphlet or food menu within which Obiageli had cleverly hidden a single United States one-hundred-dollar bill. My return correspondence was a fat envelope, stuffed with my examination scribblings, its top quadrant covered by varied stamps with pictures of birds, hippopotamuses and prancing leopards.

Obiageli secured a place for me at Cornell with these papers. It is to her I owe my career in physics and my life in the United States. My only contribution to this effort was to take two tests. For the first, I had to go to a test centre in Bodija associated with the United States embassy, where I was presented with the GRE Advanced Mathematics examination, a test that most postgraduate maths students or practising mathematicians have taken at least once in their life. Getting me to the centre took several attempts. I presumed this had been coordinated between my father and Obiageli, but I would learn in the month I lived with my brother when I first got to the United States that he had served as an intermediary between my father and my sister, delivering messages in a series of long-distance and transatlantic phone calls. The test itself I completed at such speed that the invigilator smiled sympathetically and said, 'Don't worry, you can try again. People are surprised by how hard it is and give up.'

My father, who had stayed to bring me home, spoke to the invigilator. 'Next time he will be better prepared.'

I did not return to the test centre. The results were impressive enough that I was sent a large envelope by Cornell University. Inside was a much more difficult, subtle and involved mathematics test on seven printed pages. Enclosed was an envelope to return the answers. There were no instructions or time constraints like in the first test. I enjoyed working on the problems, especially the subtle problem that required imaginary numbers.

I first stumbled on imaginary numbers when I wrestled with cubic equations at King's College. I did not then know what they were and assumed I had made some fundamental error. My excursions in mathematics were self-organised and there was no one I could consult on a development that seemed to veer off the correct path. Our maths teacher, a man named Odukwe, like the Diviner, was a dandy, more interested in the image of his brilliance than in rigorous mathematics. I soon learnt that he avoided any problems he could not immediately solve. His favourite trick, usually reserved for the presence of a female member of the teaching staff, was to stand before the class and quickly do divisions to applause. 'Kobidi, you are brilliant, but you are lazy!' he would say, looking up from my otherwise perfect answer sheet. 'Set out your proofs. What is 755? Seven hundred and fifty-five chickens? The answer is 755 km!' And I would lose two marks.

The import of these imaginary numbers would later become very clear to me. In King's College I had abandoned the first imaginary numbers I saw because they suggested an absurdity – to get an answer I had to assume there was such a thing as negative area, a negative space. I continued only when I reflected on this idea not being any more absurd than the idea of negative numbers. What could a negative number like minus four possibly mean? How could it possibly exist? How could there be a number of things that was less than zero things?

The true lesson imaginary numbers would teach me was to separate myself and my mathematics from the world of sensible objects around me; by abandoning the connection to so-called reality, we free ourselves to discover reality's true nature. But in those days all I thought I was doing was discovering new geometries. I did not yet connect the maths I was doing to the nature of our universe.

After a couple of weeks, I placed my work in the envelope and left it on my desk. A day later, my father picked it up and mailed it. Nnamdi told me that when that envelope got to Cornell I was immediately offered a place of study. I wondered how they were sure people did not cheat on the exam. When I asked Obiageli, she told me that one of the professors had said: 'We don't worry about cheating. People that can score this high are so vanishingly small that the chances they would be helping a boy in Nigeria are negligible.' She said he had added, laughing, 'We've done the math. And we are good at math,' he said.

Nnamdi was at John F. Kennedy Airport in New York to pick me up. He held up a cardboard sign with my name on it. I couldn't decide if he was holding it up as a joke. I saw the sign before I saw him. He shook my hand and then hugged me. He hailed a yellow taxi and placed my small suitcase in the boot. We did not speak much in the cab. Instead of going to Nnamdi's apartment in Manhattan, the cab took us to a barber shop in Brooklyn. My hair was matted and overgrown and had to be shaved off. Nnamdi stood in front of the barber's mirror looking at himself as he gave instructions. 'Just take it all off. Everything.' The buzzer mowed, and in the chair, I barely moved.

'Dude, what happened to you?' Nnamdi asked. I started laughing. Nnamdi's eyes dropped from the mirror. They were fixed on me and tears welled in them.

I lived with Nnamdi for about a month before he drove me up to Cornell and I started my career as a student and then a professor of physics. It was a month in my life that I remember fondly. Nnamdi had a job at an investment bank that was intense and consumed him for hours. He would, however, return from the office at about 6 p.m. Because it was summer and we were in the northern hemisphere, the day was still light and we would go out to Central Park. On the lawns, we would watch carefree people throwing frisbees. Nnamdi would buy ice cream in cones as we headed deeper into the park. He would put his arm around my shoulders. I remember a day when we listened to a reggae concert in the park until it was quite dark. There were a few women and men with us. The music and the air infused with the smell of burning weed was so carefree and easy that I danced when the girl with the palest skin asked me to. At the end of the night, about six of us sat in a cramped Jamaican restaurant eating oxtail and rice and beans. Nnamdi kept leaning over to play with the tresses of a tall Dominican woman. I drank many beers. Nnamdi and I walked into his Midtown flat at about 1 a.m.

'I am so glad to be out of there,' I said, leaning back into Nnamdi's white couch. 'The place is full of crazy people.' I smiled. Nnamdi roared with laughter. There was no way for him to know if I was speaking of the house in Ibadan or the restaurant on Fourteenth Street.

Pupil and Teacher

There is a picture, taken from the back of a seminar room in Clark Hall, in which I stand next to a blackboard full of equations, holding a pointer in both hands. At the bottom of the board, barely legible to any but the most diligent observer, is the Kobidi Equation. My wife would later have this photograph framed and hung in the foyer of our home. In the foreground are some members of the faculty and a few of the postgrads. Professor Rayburn is seated, and next to him is Professor Orhan, who would eventually win a Nobel for some of his earlier work on cold matter physics. Standing between them and cast in profile is Philip Bousquet, who has a hand on his chin. His blond hair falls to his shoulders; some of it tucked behind his ears.

I do not remember exactly at what point the photograph was taken at my thesis defence. The seminar room had been packed. This is unusual for a thesis defence. The picture must have been taken when the room had cleared and only some of the most intelligent and engaged were still making conversation, probing and questioning. The three papers that I had published had caused a great sensation in the field, and my thesis, laying out more fully and methodically the Kobidi Equation, would mark my highest point in the esteem of my peers. There was surprise, excitement and promise in the air. 'You will note that my husband is the only black man in the room,' my wife would comment as the guest at our home peered closer at the framed picture, as if to clarify. The comment usually caused me to wince.

That afternoon, I stood at the front of that seminar room, took a deep breath, and started to speak. For the first time in three years I spoke complete sentences, the words falling out of me like they had been pushed off a ledge. This was the end of the three years in which I had vowed to limit the boundaries of my ignorance and the attractions of the superfluous by limiting my responses to 'yes', 'no' and 'undetermined'. I wanted to hear; so I stopped talking.

I had done mathematics – rigorous and systematic but essentially so fundamental as to be almost rudimentary – in Ibadan. It was this work that had opened a new path for physics. Even at this moment, when I understood this new dialect, its sentences and grammar, I was still lost on what it all meant. The meaning of what I heard in that period when my lips were – in a manner – sealed, would come to me magically through my last theorem.

What we know is eclipsed by what we feel. That insight is why I am talking to you. It is also why you are listening to me.

Nothing means anything except in connection with something else. It is this relationship – this communication between you and I – that imbues what I have written with meaning. This is what makes this story exist. And when this story starts to exist it starts to matter.

8

BLACK BODY RADIATION

A Dorm on Campus

A black body is an idealised object that absorbs all the light thrown at it, internalising all this radiation, burying it all within itself. However, heated sufficiently, all bodies, including black bodies, give off light.

At the end of my first summer in the United States, I got into the passenger seat of a rented red Mazda and my brother drove me to Ithaca. As we rode up the turnpike, I looked out of the car's windows at the expansive highway covered in fresh black asphalt which broke at regular intervals, like the ridges between the bones of a spine. The regularity of the ridges resounded in my ears in soft thuds, rhythmically rising from the car's wheels. The periodic clacking sound came in pairs – after the front wheels drove over the ridges and then again after the rear.

The highway was lined with tall trees, with large trunks and leaves in startling hues of orange and yellow, and the hills seemed to stretch beyond the limits of my sight, out into the horizon. These distances protected me from the furry anxiety that would overwhelm me whenever I contemplated this trip to a new place, a place where I was supposed, again, to attempt becoming a scholar.

After about three hours' driving through the gentle slopes and curves of the highway, we got off at an exit and drove through a maze of twisting roads. We were close to Ithaca. The trees changed. They had thinner, bent trunks and grew close together, their leaves various shades of green. Several streams seemed to flow in the bushes beside the twisted roads – or perhaps it was the same one, meandering. It seemed the road led us deeper into a valley, as if burrowing into the earth. The thought caused me to start to lose my calm and I began to worry. Then the car climbed a hill. We arrived at Cornell.

Nnamdi raised the lid of the trunk. He lifted the old-fashioned brown valise, the only piece of luggage I had carried with me from Ibadan to the United States. The suitcase had been my mother's. In it was everything I possessed. A few shirts and trousers – including two pairs of blue jeans – and the sweatshirts and sweaters my brother had purchased for me at a discount store in Lower Manhattan.

Although it was just the beginning of autumn, I was wearing the winter coat bought for me at the same establishment. I would use that blue winter coat for the six years I spent at Cornell, before I met my wife Heidi, and she bought me new coats.

My brother walked me up to Telluride House and spoke to a woman standing on the porch with a folder in her hand. She had red hair and a pale complexion that was slightly lighter than the yellowish colour of condensed milk. Her colour did not have the same consistency though, and varied with pinkish hues that blotched here and there in an indiscernible pattern. She held her hand out to me and said her name. I shook her hand.

'Welcome, I will give you a tour and help you settle in,' she said. 'How much do you know about Telluride?'

I did not speak. She glanced at my brother and then continued.

'Well, the house was founded by an American industrialist to provide a haven dedicated to intellectual inquiry. Residents are provided with free room and board. Your housemates are a very select group of Cornell University's brightest students, and occasionally a few postdoctoral researchers and faculty members.'

'It's quite an institution,' my brother said. 'They have a chef to make meals and smart people for you to talk to.'

Then my brother hugged me, pulling my back into him. He let me go and smiled. 'Hang in there, Ezeani. You'll do great!' Then he turned and I heard him making his way back to the red Mazda.

I held the valise in my right hand as I was ushered into Telluride House and introduced to some students in the large wood-lined hall by the red-haired woman. I had forgotten her name. I did not speak, even though I saw that her eyes sought to compel me, in a kind way, to engage with her words.

She led me upstairs to a bedroom in which I was to spend my first three years at Cornell; the years I spent earning my undergraduate degree. I had failed to meet any of the require-ments for my mathematics degree at the University of Ibadan. It was here at Cornell that I would learn physics, thanks to an American innovation which allowed me to take up a course of study with a major in Pure Mathematics and a minor in Mathematical Physics.

A Class of My Own

I pulled out my clothes, hung the shirts, sweater and jacket in the closet, and placed the jeans in the top drawer of the dresser. I zipped closed the empty valise, not bothering to latch the two faux-leather buckles, and pushed it under what was now my bed. I took off my shoes and lay down. The room was quiet, but through the walls and floor I could hear the house, a muffled, jumbled hum of vibrating wood and human conversation. The voices were in the upper register I would associate with America and that part of its population that was described, misleadingly, and almost universally, unthinkingly, as white. Besides the voices, there was also the sound of clanging metal and plates. The house was eating. I closed my eyes.

I woke up confused the next morning. It took a moment before I realised where I was. I went into the bathroom and took a shower, making sure that I brushed my teeth. These were habits that Nnamdi had made me practise, over and over, in the month I stayed with him. 'Don't think about it. Just do it. The moment you get up,' he said. 'It's not something you want to start thinking about, debating the merits or options. Just do it immediately, automatically.' I was learning again the things my mother had taught me when I was a child, after I had been consecrated a high chief I was already a titled man before I learnt to brush my teeth.

When I came down the stairs the woman with red hair was seated in the hall waiting for me. 'Good morning, Ezeani,' she said. 'First port of call today is an orientation tour of the faculty building, and then, registration.'

The red-haired woman registered me for my first courses at Cornell, filling out several forms and walking me through many buildings and, occasionally, briefly consulting me with respect to my preferences. She seemed eager for me to provide even the barest response.

I was struck by the beauty and – there is no other word for it – majesty of the university campus. It was on a scale comprehensible to me from my childhood at the campus at the University of Ibadan, but the nature of these buildings was of a very different sort. The buildings at Cornell I associated with the castles and forts in fairy tales, primarily set in Europe's Middle Ages, which I had read in illustrated books from the lower bookshelves in my father's study. But even more than the buildings and the curved avenues, the large lawns and the great trees in their froth of rich reds, oranges and purples made the deepest impression on me.

The routine that the red-haired woman established in that first week carried me through my first years at Cornell. I varied little from it and then only in minor ways. For instance, between lectures, I developed the habit of immediately going into the postgraduate student lounge in the physics faculty building. The room was usually empty. Most of the other students preferred the cafeteria where they could eat or the library where they could study. This room, with its comfortable chairs and television, I would later learn, was not accessible to undergraduates, and I had access because of some special dispensation or some confusion regarding my status.

In my first week at the university, the red-haired woman guided me to all my classes. She would walk me to the door of the classroom and wait outside for the hour or so it generally took for most of my lectures. I say she would wait outside, but now, when I reflect on this, I cannot be certain that this is what happened. All I can say is that when I came out of the classroom, she would be in the hall waiting for me. Thinking on it, it seems logical to assume that she took up some other tasks and only returned to the hallway just before the class concluded.

In that first week, the red-haired woman attended one lecture with me – the survey course introducing advanced undergraduates and graduate students to physics. This lecture was held in what felt like a small auditorium, with seating that sloped from the back to the front. A grey-haired man, clean shaven and wearing a fading red cardigan over a white shirt, walked up on to the podium and started teaching me physics.

'The first clue that we do not live in the conventional world of objects, of tables and chairs, but in a quantum world, was when it was discovered that heated black bodies do not throw off light progressively as heat is increased, but only in narrowly defined bands. As if in quanta,' Professor Rayburn said, pointing at the board.

The Burdens of History

As the semester progressed, the days started to cool. The bright orange and yellow in the trees dulled to an unpleasant damp brown. My routine was set. I walked from Telluride House to my classes in the morning. In between classes, I worked in the postgraduate lounge. Then in the early evening, I walked back to Telluride where I had dinner with my housemates, the sound of voices, silverware and crockery floating in the air.

As the days got colder, I wore my blue polytetrafluoroethylene coat everywhere. The heating came on in all the rooms – the lecture rooms, the dining rooms, and the postgraduate lounge. The heat carried smells. As the rooms started to warm, and the pings that came to my ears from expanding coils in the heaters became so regular that I wondered if there was a pattern, my nose was filled with this indistinct stench. I had difficulty pinning down this odour that travelled with the heat, generated by gas furnaces and electric filaments, which warmed the ambient temperature of these enclosed spaces.

One evening, I sat in the hall at Telluride, in a large armchair in front of the unused fireplace. A few of my housemates were in the large room talking. A woman with bright, monolid eyes and jet-black hair turned to me.

'What do you think, Ezeani?' she asked.

I looked up at her and smiled quickly, then looked into the book on my lap, looked up and smiled again.

'Ezeani, you must be one of the strangest residents at Telluride. And there have been a lot of strange residents here,' she said.

The black-haired woman turned to look at the porter as he walked across the length of the room and put an envelope into my hand. It was a long, white rectangle and in one corner were small stamps, cropped United States flags. In the other corner, in my sister's handwriting, was her name and an address in Connecticut. I placed the letter in the inner pocket of my polytetrafluoroethylene coat.

Later, when I sat down to dinner, the conversation at the table was principally about music. I barely listened. On my plate was chicken with a brownish sauce that contained mushrooms. One of my housemates asked: 'Does anyone have a view on the connection between meaning in music and mathematics?'

'Oh, please don't make me groan,' Philip Bousquet responded, as he groaned. The red-haired woman had gone out of her way to introduce me to Philip, pointing out to me that he was the only other student in Telluride House that year studying physics. He was a year ahead of me, and in my first year we had no lectures or classes together. Although he had been friendly when we were introduced, smiling as he tucked his long yellow hair behind his ear, he had made no other effort to engage me. This did not distinguish him from many of my other housemates, few of whom made any attempt to speak to me. A dedicated number, however, would reach out to me, making repeated efforts to draw me into conversation, bring me along on excursions and invite me on errands. Most of these efforts were unsuccessful; the only reward for them, and response from me, was often just a small, shy smile.

Occasionally, however, my response was grander. I would agree to go down with a group of three for a pizza run to Collegetown. Standing in a small pizza parlour with cardboard boxes heaped in my arms so the stack rose to my chest, I would smile and laugh as the yellow-haired female housemate paying for the pizza joked and said, 'Strong, dark and handsome, Kobidi. You're the whole package.' And I would look over and see that the large man behind the counter reaching out his hairy arm to give change was smiling too.

After dinner, I sat up in my room and opened my sister's letter. I read it, refolded and placed it back in the inner pocket of my coat. Then I hung my coat on the back of the reading desk chair.

I woke up from a nightmare with the taste of sea salt in my mouth, struggling to see through the darkness. My pillow was wet. As my eyes adjusted, I saw Anyanwu's face above mine, his eyes cruel and full of rage. I screamed. As the sound left my mouth, Anyanwu blew a froth of saliva and herbs on my face. Its sweetness mingled with the sea salt in my mouth, and I could taste something ugly as it trickled down my throat. I felt the urge to vomit rise from my stomach. Anyanwu's large, calloused palms covered my nose and mouth. I could not breathe. His eyes burnt. He was speaking to me. And he was repeating himself. 'Don't be afraid. Don't be afraid. Don't be afraid!'

I was terrified. Then I lost consciousness.

When I was revived, I could still taste sea spray in my mouth. My ears were filled with the whistling of a storm and the sharp cracks of thunder. Then there was a strange, loud crashing sound, a large wave pounding against wooden boards. When I opened my eyes, my hands were tied by thick raffia rope to a stout pole. I was on the deck of a wooden ship. The sea rose and fell in large swells and the sky was a dark, menacing grey, flickering in patterns of dark and light silver that radiated the moonlight falling in the spaces between the large clouds. I felt iron cuffs cutting into both my ankles. I turned to look down. Then I felt the sting and the wrenching burn of a whip lashing my back.

'He will pass out again. Stop!' A voice in an upper register came to my ears. I looked through the darkness and saw a yellow-haired man in a dirt-brown wool shirt, his arms pressed into his waist in a fighting stance. In the moon's flat light, his skin was the colour of tin. 'Who are the others with you? What are you planning?'

I was surprised. He was speaking to me in Igbo.

'What did he say?' Another tin-coloured man moved from behind me and stood next to the man who had spoken in Igbo. He held a dark brown leather whip between his hands. This man's hair was red. He spoke in English.

'He is muttering gibberish. But from what he says, I think the ringleader is named Anyanwu. We need to figure out which one he is. They are wily, these Igbos. He will not answer if he is called,' the yellow-haired man said.

'Maybe we should consider throwing all the male slaves over-board. I am afraid it might have come to that,' the man with the whip said, and he started to move closer to me.

'That's a decision for the captain. But I think you are right,' the yellow-haired man said. The red-haired man paused. Then he grimaced, and started walking towards me, unfurling his whip. His eyes were filled with rage.

As he got closer, the tin-coloured flesh became the colour of a goat's bone, a pale yellow, splotched with brown. He smelt. The stench of heat, raising by degree, the ambient temperature of human flesh. Later, I would realise that it was the same smell I had come to associate with the rising heat in those enclosed rooms in winter. A smell that would permeate the late fall and winter for me in the years I lived in the United States; a smell from which I would often seek relief by going outdoors, in my blue winter coat, with its lining that unfortunately trapped odours. Odours that I carried outside, so that traces lingered in the pristine cold air.

When I woke up again it was day. My sheets were drenched with sweat and the unmistakable stench of piss. Anyanwu was seated at the foot of my bed. He was wearing nothing but his loincloth and the floppy cap with an erect eagle feather. He was staring at me. 'Don't be afraid,' he said. And placed his hand on my ankle.

'Shut up!' I yelled. 'I am no longer a child you can terrify with nightmares.'

Anyanwu started giggling. He laughed, gently at first, then so loudly that his head shook, and tears came to his eyes. Then he got up, walked to the door, opened it and walked through. 'It is myself I blame,' he said as he slammed it.

The Pumpkin Festival

'The challenge of racial justice is real. Even in the most liberal places there is significant struggle. Think about it, there are only two black students in Telluride House! Virtually everyone else is white,' the jet-black-haired girl said, twirling pasta on her plate with a fork.

The custom at Telluride House was for the residents to share dinner in the building's basement dining hall. These dinners were a key part of the function of the house, which – as I understood it – was to permit a forum for the intellectually gifted in various fields to interact, sharing ideas and building habits of the mind that would shape the future of the world. For me, it was a roof over my head. I usually sat mute at these dinners. Occasionally, I listened to what was being said. Rarely, I would contribute to the discussion in terse sentences that seemed to strike my housemates as quaint or droll, judging by the anticipatory hush that descended on those few occasions I spoke.

'There are no black people at Telluride House. I'm brown. So is Ariana,' I said.

'It's just a convenient description, Ezeani.'

'It's inaccurate. And needlessly so. Brown is a colour. Same number of letters,' I responded.

'The point Kailani is trying to make is that most of the residents at Telluride are white,' said a housemate. He was wearing a sweater that contrasted sharply with his auburn-coloured hair.

'There are no white people at Telluride,' I said, pulling my plate closer to me.

'I'm white,' the auburn-haired student said, starting to smile.

'No, you're not. You are a shade of pink,' I said.

'I'm white. It's a category we use in America, and I think pretty much around the world,' he said emphàtically. The room quietened in response to his raised voice. 'Perhaps a flush of pink when I blush,' he added in a softer tone, and then smiled.

'Have you ever seen a pig?' I asked. 'Would you ever describe that colour as white? You're the same colour.'

There was a roar at the table and then some laughter. 'Now, now, Ezeani, we don't do personal insults here,' the jet-black-haired girl said. I looked at her, my face blank. I did not understand what she meant.

'What I think Ezeani is drawing attention to is the absurdity of using the term "white" to describe any collection of *Homo sapiens*. It's neither accurate nor neutral. If you actually think of it, it's a ridiculous statement. Self-aggrandising,' Ariana said, then paused. 'Like, "We ain't people of colour like y'all. We white."' Inexplicably, there was a lot of laughter.

Later, when dessert was being served, I was recruited to join a pumpkin-hunting team by the yellow-haired girl. 'We have all the culture and sociology stuff covered but we need the hard sciences. We are asking you and Phil. One of you has to say yes!'

As it turned out, Philip and I both accepted the invitation. On the day of the hunt, we waited in the yellow-haired girl's red car for the first clue to be delivered by the hunt's organisers.

The hunt for pumpkins on the last day of October was the most fantastic diversion in my first year at Cornell. Of our party, today I'm afraid I can only remember the name of one of them. Six of us were crammed in the red convertible Volkswagen Jetta. The girl with the yellow hair turned the steering in a wide arc and we were pushed to the opposing side as the car made a loud U-turn.

It was a long, active and peculiar Saturday, ending with all six of us squeezed into a booth at an Irish pub, a gold-painted pumpkin sitting on top of the table with a long cigarette inserted in the small mouth that had been carved below make-believe eyes and nose.

We were served numerous pitchers of beer, round after round, by a dark-haired waitress wearing a green jersey and a black apron. I was light-headed, light-hearted and filled with energy and joy. Everyone at our booth was laughing or smiling.

I got up from the table and stumbled out of the door. The

night, lit by the electric lights from the rows of open bars, seemed frozen. I walked down the street and my nose filled with the sweet-sour smell of alcohol. When I got to the corner, I stopped, leaned against a brick building, and vomited. As the puke left my mouth, I tried to draw breath through my nose. The stench of my own vomit surprised me. I spat at the ground and then cleaned my mouth with the back of my hand. As I raised my head, I noticed, a couple of feet from where I stood, a man wrapped in a dark duvet, his head propped up against the building's brick wall. The man – I assumed he was homeless – stirred.

'Brother, help me out,' he called. 'It's cold, man. Anything you can spare.'

Because I had just vomited on his resting place, I reached into my coat pocket to see if I had a few loose dollars. My hands felt the envelope containing my sister's letter. Then I felt crumpled money. I had a dollar. I pulled it out, leaned in and placed it on the man's duvet.

I started walking up the hill towards my residence. I could hear the man muttering something I assumed was thanks. He seemed to be trying to get up. I quickened my pace, glancing back then facing the road in front of me. Immediately I turned the corner, the road was empty and silent; all the bustle and sound from the row of bars had died away. I heard footsteps behind me. I walked even faster, keeping my eyes fastened on the road ahead. The footfalls, sharp and hollow, continued, and though the cadence did not increase, the footsteps sounded closer, as if he was just at my back.

Finally, out of sheer terror and not courage, I summoned the will to turn around. The man stood still, his face in the shadows, less than a hand's length from mine. On his head was a yellow fedora.

'I know you are upset with me, but what happened yesterday is not my fault,' the man said.

'What happened yesterday?'

'On the ship. With the goat-bone-coloured men that flogged you and drowned our kinsmen?'

It was clear to me now that this was Anyanwu.

'That didn't happen yesterday. It happened over five hundred years ago.'

'It is still yesterday. No propitiation has been offered to me or any other of our Gods. I still stand in their stead. None has been offered. None has been accepted.' He held his head up, looking fiercely at the moon, his arms folded above his chest. The dark duvet fell to the ground. He wasn't wearing a shirt. His muscles rippled in the pale moonlight.

I laughed. Slowly at first, then with a large laughter, shaking my head.

'Ezeani! Ezeani! Ezeani! My acolyte, do not laugh with that bitterness. I know you blame me.'

'Why would I blame you?'

'Because you ask yourself: "What kind of God lets such a thing happen to his people?"'

'Good question,' I responded. 'But that's your business. And why do you keep calling me your acolyte?'

Anyanwu was silent for a moment. I started walking again. He picked up the duvet and kept pace beside me. I stuffed my hands into the pockets of my blue coat. After we had walked about a block in silence, Anyanwu started to speak.

'Ezeani, it is cold here.' He pulled the duvet tight around his shoulders. 'Since yesterday when I left your room, I have been living outside, lying on the roads, where I can see the sun and the moon. The cold though, it's another thing.'

It had certainly been many days, perhaps weeks, since

Anyanwu had passed through the door of my room. For a moment, I was confused by his reference to 'yesterday'. I was trying to make sense of the order of days and to arrange my memories in time. Then I remembered that he had also referred to a distance in time of five hundred years as 'yesterday' and decided to abandon this exploration.

'You are a God; why are you on the streets like a vagrant, asking strangers for money?' I asked.

He stopped walking. I stopped too. He turned his eyes to me. His face was strong and handsome, the features sharpened by the moonlight, but his eyes had softened. 'It is a sad thing, I acknowledge, that I, God of the Light, Preparer of the Dawn, beg humans for alms. I do it because our people abandoned us. I do it because your faith in me, Ezeani, shakes and wanes, it comes and goes. I beg because when these humans offer me alms, they acknowledge me; their tokens keep me alive.'

I turned away from Anyanwu and kept walking. He sighed as he caught up with me and placed his rough hand on my coat sleeve.

'Ezeani! My son, please wait. You ask why I call you my acolyte. It is because you are the one who will bring me fame among humans. You will make them remember my name. Perhaps, you yourself do not yet know. I see now that it was always destined to be so.' Anyanwu turned up his head and sniffed the air.

'This is a strange place. Ignorant people who celebrate the Festival of the Pumpkin Leaves at night! But I see now why the calling of your Chi, the destiny that is drawn in the palm of the right hand of your Ikenga, has brought you here.'

The street was empty, with shadows thrown by the moon through tree branches and leaves appearing like mysterious script on the pavement. Anyanwu was consumed with his words, his eyes focused on me.

'I am here to make you famous?'

'It is more than that. Your journey, Ezeani, is to the Place Beyond Knowing. You are destined to know what no other knows. And it is you, my acolyte, a man of our people, that dares to go to this place; to know this thing . . . !' He stopped talking, shook his head in wonder. There was silence.

'What is it that I will know? What is this place beyond knowing?' I asked.

'Let me do something for you. The human does for its God and the God does for its human. Let me tell you something you don't know. The journey to the Place Beyond Knowing is not a thing to be trifled with. It is a place beyond Gods, beyond understanding. No one, God, man or spirit, embarks on this journey without their kith in their right hand and their kin in their left. There is no God that guides in the Place Beyond Knowing. You must be fortified for this journey. I shall fortify you. I shall prepare you so you cannot be shaken from the bottom or accosted from the top.'

I stood in the moonlight on the hard pavement and my head dropped. I had expected him to say something concrete, useful. 'That is it?' I asked. 'That is all you have to tell me?'

'I have told you what I know. It is not everything, but it is not nothing. Look, all these other Gods go around pretending they know everything. They don't. All you have to do is listen, and you can hear their ignorance. Ezeani, at least I tell you the truth: I don't know.'

I looked up at Anyanwu and smiled. Then I chuckled. 'Don't forget the five hundred years that have passed without anyone coming, trembling in fear, to offer propitiation.'

Anyanwu stared hard at me and then struck me across the face with the back of his hand.

Heated sufficiently, all bodies, including black bodies, give off light. I held my face and stared at him.

'Do not think you are still a child that can wait for its mother to slap his hand away from a scalding pot. The journey is fraught. You will need me. Onye ka nmadu ka Chi ya.'

He turned around and started walking back down towards Collegetown.

A black body is an idealised thing. It is a creation of our minds, the perfect receptacle for all the light and radiation which is thrown at it. It does not, even in defence, reflect any radiation back at the world. The black body absorbs all. But black bodies, when heated enough, like all other bodies, will give off light.

I watched him till he was almost out of sight, then I turned and walked towards Telluride. As I walked, the streets seemed to come alive. There were shouts and hoots. Masked men and women appeared on the sidewalks holding up lanterns shaped like manic pumpkins. They walked in small groups of three or four. There was a lot of laughter, and I could hear the sound of empty beer bottles.

I had no fear of them. Not until I was close to Edgemoor Lane and two spirits, shaped, in height and build, like plump children, jumped out of a bush and shouted, 'Trick or treat!'

9

SUN GODS AND WHITE GIRLS

Obiageli's Red-Haired Friend

My first winter in Ithaca was unseasonably warm. Although the big trees were stripped of leaves and the morning's cold frost clung like icicles to the blades of grass on the front lawn, by afternoon, the frost had thawed and the heads of grass drooped, like those of disappointed people. There would be hardly any snow until later, when a storm in late January would acquaint me, for the first time, with piles of fluffy, frozen water.

But in December the house was quiet. There were no voice whispers, and the boards did not creak. At the start of the Christmas break, most of my housemates had left to spend their holidays with family. I and a classmate who had arrived mid-semester from Mainland China were the only students left in the house. In my hand I had Paul Dirac's *The Principles of Quantum Mechanics*, and my back was against the wall, supported by a pillow. That book is a treasure, and I took it everywhere in those years, reading it in quiet moments as the world unfurled before me.

There was a knock on the door, so gentle that at first I wondered if I had imagined it. Then there was another knock, louder and more insistent. I made a noise. The door swung open slightly. I expected to see the round, shiny face of my

Chinese housemate. A moment seemed to pass and then the red-haired woman's head appeared between the door and its wooden frame.

'Hi, Ezeani,' she said. 'Can I come in?'

I grunted and she walked into the room. She moved to the edge of the bed, leaned, and then sat beside my feet. She had a folder in her hand.

'What do you want to do for the holiday, Ezeani? Would you like to go to stay with your brother or sister?'

'I prefer to stay here,' I replied.

My sister had written several times offering to host me for Thanksgiving and then the Christmas break. She had also offered to arrange for my brother to pick me up at Cornell and drive to her home in Connecticut. She suggested we could spend the holiday together as a family. I had not replied to her letters. I had also ignored my brother's half-hearted efforts to get in touch. His last message, scrawled on a note left in my mail slot, had provided me with a deadline, after which, he said, he would be committed to a trip to San Juan, Puerto Rico, with a girl-friend: 'A fine specimen.' The word 'fine' was underlined. Reading the note, my mind went, not to my brother, but, in a process I have noticed several times in my life, linking parts of what appear to be my past and my future in inexplicable ways, to the image of Chijioke wearing jeans with a tape in her back pocket as she walked out of my dorm room. I was momentarily unsettled by this image. Then I wondered at the telephonic exchange between my brother and the incidental scribe, prob-ably a secretary in the Physics Department, which had resulted in this absurd note.

'Your sister is in Ithaca to see you. She would like you to come and meet her. She's at my place. I can drive you over if you'd like,' the red-haired woman said.

'I don't want to,' I quickly replied, barely waiting for her to finish speaking.

'She just wants to make sure you are fine. Why don't you sit with her for a few minutes and then decide what you want to do? She's come a long way.'

'What do you know about my sister?' I asked. I could see from the face of the red-haired woman that she was surprised at the vehemence of my response.

'Well, I know she is a great physician. We were classmates in medical school at the University of Connecticut. She asked me to look out for you, help you adjust to Cornell. As I told you, I work as a physician here in Student Support Services. Taking care of the needs of students with special challenges is part of my job. Because your sister asked, I've paid attention to you. She has given me some background and she's really been following up.'

I did not respond. I turned to my book. She did not move. After I had read a few pages, I realised I was distracted by her presence. I looked up and fixed my gaze on her face. Her skin appeared to have an almost translucent quality that was marred by an uncoordinated smear of pink and light red on her forehead and cheeks, and down her neck.

'Please tell my sister that I don't want to see her,' I said.

She didn't seem perturbed by my statement. Her eyes were fixed on me in a sympathetic gaze. 'Can you, perhaps, give me a reason for your decision? It would be helpful if I could tell her.'

'It is really simple. She killed my mother. She killed her with her selfishness and hate,' I responded. I turned my eyes to my book.

A few moments passed and then she said: 'Why do you believe this?'

'Because my father told me the cruel things she said to our mother.'

'Did your father suggest your sister is responsible for your mother's death?'

'Actually, it's the opposite. He says I should let it go. He makes excuses for her. He even asks me to forgive her for the cold, brutal way she treats him.'

'Perhaps you might want to hear what your sister has to say? Ask her about it yourself?'

'There is no need for that! I, Ezeani Kobidi, will NEVER, NEVER, NEVER, NEVER, NEVER forgive her!!' I saw a hint of fear in her eyes as I stood up and let this torrent of words pour from my mouth, then beat my right fist against my chest, the way I had seen Anyanwu do on a moonlit night.

The red-haired woman stood up and hurried out of the room. When she was on the other side of the door frame she said, 'I will let her know,' before she closed the door behind her.

The Girls in Their Summer Dresses

Summer surprised me. I was still wearing my blue coat. The sun came out and it felt like the long nights of winter had made our local star, the sun, reluctant to set. It would linger in the sky till long into the night, the result of the Earth's tilt and Ithaca's place in the further longitudes of the northern hemisphere.

As the sun tarried, the coats, sweaters and shirts fell off. There was exposed skin everywhere. It seemed as if I were surrounded by different people. The patchy, pale, vaguely pink skins of most of my housemates and the students and faculty of our campus seemed, almost in an instant, to disappear, replaced by skins of varying shades of bronze. I was mesmerised by this sudden, unexpected transformation. The shades were intoxicating in their sheer variety. From the light copper tones – which for some reason reminded me of the hue of ripening cashew fruits – to the deep sepia browns that I came to associate with raven- and brown-haired girls, but which, in surprising regularity, one saw matched up with the blondest of the yellow-haired girls.

On my way to class I would walk through the grass squares and quads of campus and watch my fellow students in tank tops, bikinis and athletic shorts, throwing frisbees and sunbathing on the lawns. There was so much naked flesh on display. So much beautiful, bare, bronzed skin covering the length of legs, the curvature of breasts, the span of hips, all so fetchingly and unselfconsciously displayed. The women would sometimes lie face down with the straps of their bikini tops undone so the sun would brown otherwise-covered strips of pale skin.

I must mention here that, in this period of my life, I was a virgin. I would have remained one until I had intercourse with my wife Heidi if not for a memorable encounter engineered by my brother Nnamdi just before I started graduate school. But virgins are not lacking in sexual desire.

The abundant summer skin stoked my lust. It was a sudden, persistent flame that would rage until the autumn days arrived, shortening the path of the sun over the skies of Ithaca. The pale skins returned, and my desire died, shrivelling like a seedling blocked from the light.

But while the sun was in the superior position, the sight of browned skin on the blonde-haired beauties was almost more than I could bear.

I would walk through the quad on my way to my lectures, glancing quickly from side to side, recording the bountiful sights like a trespasser hoarding at a sumptuous buffet. Once I made it into the faculty building, I would go straight to the bathroom, enter the furthest stall, pull out a small tub of Vaseline from the pocket of my blue coat and, sitting down, quickly masturbate. I would be forced to repeat this action several times a day, as often as was required to sate the desire that rose up in me, often for inexplicable reasons: the way a pen was held to the mouth, the slope of a dangled elbow, and once, simply, the sight of deep dimples on a round, bronzed face.

An Assault on the Senses

One evening in the June of my third year at Cornell, I was at my desk reading Dirac when Philip Bousquet knocked on my door. He popped in his head and told me a few housemates, himself included, were going down to a new nightclub in Collegetown. 'Jenny says it's hot or dope or something. Let's go.' I was rarely invited on these outings, and when invited, rarely accepted. But on this occasion, because it was Philip who was extending the invitation, I decided I would not decline.

'Let's go!' I said and stood up, walked over to my bed, sat down again, and started putting on my shoes.

Philip was looking at the papers on my desk, turning the pages very slowly. I lay my shoulders back on the bed and watched him. He stood for a little longer, then he sat down on the chair and continued reading my notes. We must have been in this state for almost half an hour when a blonde-haired girl pushed her head into my room. 'Come on, guys!' she said. 'We've been waiting downstairs.'

'Yeah, let's go,' Philip said distractedly, making no effort to leave the desk. The blonde-haired girl stepped into the room and grabbed Philip's hand, drawing him up from the desk. 'OK, OK, Jenny,' Philip said as he stood.

I stood up too. 'I am ready to go,' I announced.

'You might think of changing that shirt, Ezeani,' Philip said.

'We will wait for you downstairs,' Jenny said. 'You have ten minutes, Ezeani. I mean it.'

A few minutes later I walked down and met Jenny and Philip on the porch. She was holding the back of Philip's long blond hair between her fingers. I was wearing the same striped shirt but I had put on my blue polytetrafluoroethylene coat, with the zipper pulled to the top.

'Dude, you'll boil in that,' Philip said, shaking his head. They turned and we all started walking towards Collegetown.

The building's interior pulsated with neon lights flicking in a kaleidoscope of blue, red and yellow. Hip hop music played at a deafening volume. Jenny led our party through the crowd to the border of the dance floor. Jenny and Philip started dancing, with Philip's arms around her waist. I placed my hands into the pockets of my coat.

'Drinks?' Jenny asked when the music changed. I could barely hear her. Philip whispered something in her ear, and she walked off. Philip and I stood in silence on the edge of the dance floor. I was sweating inside my coat.

After what seemed an inordinately long time, Jenny returned, beckoning us, her arms raised over her head. We started moving towards her. When we met, she leaned into me and shouted: 'There is a Nigerian prince in the VIP! He wants us to join him.'

As we approached the red velvet ropes marking off the VIP area, I saw a group of about eight people lounging on large white couches. Three of them were men. The rest were tall, beautiful women. The bouncer stopped Jenny, holding out his hand. 'We were invited by the prince,' Jenny said. A tall man in a tight-fitting black shirt that stretched to outline well-formed muscles, flanked on either side by two women, walked up and whispered in the bouncer's ear. He lifted the red velvet rope.

Jenny stuck out her hand. The prince took it and then pulled her into a hug. Jenny giggled. The shafts of light passed rhythmically across his face.

'Sit down, guys,' the prince said, turning to me and Philip. 'Fix them a drink, will ya?' he said to a waitress. He had refocused his attention on Jenny, holding her hand as she sat on the couch beside him. Philip squirmed. The waitress asked us if we wanted vodka and cranberry juice. I said yes. Philip did not speak. She poured out two drinks. I lifted the glass to my mouth and drained it. Philip's drink sat on the table. I followed Philip's eyes. Jenny was leaning into the Nigerian prince, speaking into his ear, with his arm around her shoulders. I picked up Philip's glass and emptied that too.

Suddenly, the music changed; the new song caused a collective shout of joy to fill the club. The dance floor filled up, spilling over. Everyone in the VIP, with the exclusion of Philip and me, stood up and started to dance. The Nigerian prince jumped with both feet onto the white couch and then leaned down with his right arm to guide Jenny up. The bouncer started walking towards them. 'Prince! Prince! You can't dance on the couch.'

'Yo, chill out, cracker! I can do whatever the fuck I want! I am a God!' and then he beat his right fist on his chest. One of the prince's companions moved to intercept the bouncer. I poured out another glass of the vodka and drained it. I must have blacked out. My clock stopped.

The next thing that I do recall, appearing to me quickly and clearly, was Jenny, Philip, two of the blondes from the club and the Nigerian prince riding with me in an elevator. We were packed tightly in the space.

I was saying: 'What we know in physics is that everything is, in a very important sense, a type of illusion. An illusion that only appears to be real because we observe and interact with it in a certain way.' I was speaking loudly, my hand moving around animatedly. The two blondes from the club were giggling.

'Is what my subject says true?' the Nigerian prince asked, turning to Philip. I looked at the prince closely. His handsome face, well-built frame, the ripped chest muscles visible under his tight black shirt. I recognised him. He couldn't fool me. It was Anyanwu. He was slightly younger, and his facial hair was styled differently, but it was definitively Anyanwu. I had seen him in that disguise before, at the square in Collegetown.

Before Philip, who was just looking up, could respond, I said vehemently: 'I am not your subject. The Igbos know no kings. I know this guy. He is a fraud.'

'He is right!' Anyanwu agreed, laughing. 'The Igbos know no kings. I am his God!' he said, and smiled, then laughed again.

Jenny, who was leaning into him, laughed too. 'Not that again, Sun! That's how we got kicked out of the club,' she said and patted his chest. Anyanwu looked at me and winked. Then he pulled up Jenny's chin and kissed her. I turned away. Philip turned his eyes to the floor.

I don't remember how we got into the apartment. I remember that we – Jenny, Philip, two of the blondes from the club and Anyanwu – sat in a circle. Anyanwu was passing around some pills placed on a white saucer. 'That's one interpretation of quantum mechanics,' Philip said. 'An observer-determined universe.' Anyanwu was rolling a large joint. He licked the ends and then put it to his lips, struck a match and lit it, inhaling the smoke and then blowing it out through his nose.

'It all just sounds like crazy talk,' one of the club blondes said.

'Yeah, it is crazy,' Philip agreed. 'No one knows what it really means. I mean, not even our professors.' His eyes were glazed. 'Do you guys know Richard Feynman?' he asked.

'He was a brilliant physicist who won a Nobel Prize. He used to be a professor at Cornell and, while here, a proud resident of Telluride,' Jenny responded, speaking loudly so everyone could hear.

'Feynman said, and I quote,' Philip continued in a quiet voice, barely looking at Jenny, '"Do not keep saying to yourself, if you can possibly avoid it, 'But how can it be like that?' because you will get 'down the drain', into a blind alley from which nobody has yet escaped. Nobody knows how it can be like that."'

As he finished reciting, the joint was passed to Philip. He took a long draw and then stood up and passed it to me. 'Take something like the atoms that make up this couch,' Philip said animatedly, pointing to the long couch against the wall. 'In high school we are taught that it's made of atoms whose electrons orbit a nucleus, like little planets around a sun. But that's not true at all. An electron is not a "thing". We only fix a position for them when we look. Like they stop being ghosts, materialise and pose for a photograph when they are being observed.'

I held the joint. For a moment, I considered taking a puff as I had seen the others do, primarily to satisfy my curiosity, but almost immediately decided that it was safer to abstain since I was not certain of its contents and the possible effects on me. 'Do you know the key equations of quantum mechanics?' I asked, passing on the joint and turning to Philip. Philip nodded slowly. 'Then you know what it means. You just don't want to believe it,' I said.

The acrid odour of the weed sat in the air like a soft cushion on our heads. 'If you want to visualise it,' Philip continued, ignoring me, 'if an atom is the size of New York City then an electron is the size of an apple. Yet the electron's probability waves occupy the entire space of the city. The probability waves are not real, but they are real enough for you to sit.' He pointed at the couch.

'Real enough for you to fuck,' a voice added.

'What did you say?' Jenny asked, turning to me.

'I didn't say anything,' I responded.

'I said, "real enough to fuck",' Anyanwu shouted. 'Come on, blonde girl, fuck me like you're Jamaican,' he added and laughed, cupping Jenny's buttocks in his large hands.

'Sure you'll survive it?' she said, laughing. Then Anyanwu's lips spread in a lascivious grin.

When I woke up, I was lying on the couch in the living room, the taste of sour alcohol in my mouth. My head hummed. Everyone else was gone. I looked up at the ceiling. It had recessed lighting, cut into small evenly spaced holes. I groaned. Then I was silent. The sound of other groans came to my ear. From behind a wall there was the clear sound of a woman moaning. The moans increased in intensity until they morphed into a repeating squeal and then little screams of pleasure. Then the sound reached a crescendo and stopped. I lay still. I was trying to understand where I was. I was still wearing my blue coat, but the zipper was undone to the middle of my chest. There was a carpet on the floor. I could see now that the furniture looked expensive. The woman on the other side of the wall started moaning again. I think I lay on the couch for another hour. I just stared at the ceiling and when this started to bore me, I did some mathematics in my head.

The door of the bedroom opened. Jenny walked out wrapped in white bedsheets. 'Hi Ezeani,' she said as she moved past me towards the refrigerator in the open-plan kitchen. The skin of her back was a deep bronzed brown. 'Want something to drink?' she asked as she threw open the refrigerator door.

'Kedu, Ezeani! Did you sleep well?' I was being addressed in Igbo. I turned towards the bedroom door. The well-built man from the nightclub was leaning against the frame in a white terrycloth robe. In daylight I was certain it was Anyanwu.

'Yo, Jenny. Grab me some apple juice. Do you know how to fry eggs?' he said. Then he turned, smiled and winked at me.

In the early afternoon, after we had shared a Chinese takeout lunch at the table in the large, expensive apartment – I sitting on the green high-backed chair and Jenny sitting in Anyanwu's lap – he walked us outside to wait for the taxicab. The apartment complex was one of the few modern apartment buildings in Ithaca. Most of the residents of the town lived in lovely wooden houses scattered in its hills and valleys. The kind of home I would later live in with my wife Heidi. This apartment complex looked like my brother's apartment building in Manhattan, designed for high-earning professionals, probably without children.

I was surprised Anyanwu had insisted on a taxi. They were expensive. I walked or took the bus everywhere. When the white taxicab arrived, Anyanwu pulled money out of the pocket of his bathroom robe and counted out a few bills. He leaned into the driver's door. 'Keep the change,' he said.

Burying the Staff

In the middle of that week, I received a note in the slot of my mail cubby at the physics faculty. It simply said: *Come my flat.*

This afternoon. An address was scrawled below. I knew immediately who had written it.

Anyanwu met me at the door wearing a grey tracksuit and sneakers. 'My man, welcome,' he said as he opened the door. He looked younger, fresh, like a slightly older undergraduate. Now we were alone he spoke only in Igbo. 'What can I get you to drink? I don't have palm wine. I haven't yet figured out the way to get it fresh here,' he added.

I was looking around the apartment as he spoke. It was such an expensive place. Anyanwu must have caught the questions in my eyes. 'I know what you are wondering. Let me be honest with you. I owe you everything. You know Odukwe the Diviner is dead? If it was not for you, I would not exist. Although you insult and mock me, it is your belief in me that has kept me. And I am grateful. Let us sit down.'

He motioned to the large grey couch that ran almost the length of the living room, facing a large television. On the screen, people in suits were moving around in a large room preparing for some event to start. The sound was muted. A chyron at the bottom of the screen read: **C-Span: Supreme Court Confirmation Hearing for Judge Thomas.**

Anyanwu walked towards the open-plan kitchen and came back with two short glasses and a bottle of schnapps. He filled each glass, picked one up and downed it in a gulp.

'Ezeani! Ezeani! Ezeani!' he hailed me. 'The One They Call the Friend of a God; the Quiet Whisperer that Sings the Truth; the Tough Meat that Fills the Mouth!' He cleared his throat, picked up the second glass and drank. Then Anyanwu sat down, rubbed his palms on his knees and turned to me. 'I know that you are on a great mission. Let nothing trip you from below or accost you from above. I, Anyanwu, the Sun God Whose Smile Mocks Darkness, I will be here to prepare you, to fortify you on this journey. But I do not want to go to this place. I am happy where I am.' He poured out two full glasses and offered one to me. I drank.

'You are a funny God,' I said. 'A Sun God who prefers to dwell on Earth.'

'In the United States,' he said and smiled. 'I am beginning to learn things about this place. When a God has few worshippers, you have space to explore the world and yourself. There are few believers left to judge you, to disapprove of your conduct.'

Anyanwu stopped talking and again filled both glasses with liquor. He picked one up and downed it in a single gulp. I picked up the other and drank as well.

'Let me answer the question in your eyes. How did I get this beautiful abode?' he started.

'If you are a God, why can't you just conjure it?' I said, smiling.

'Because of the Law of Conservation of Matter,' Anyanwu responded. I looked up at him, startled. Then I heard Anyanwu's laughter, loud and rising. 'Ezeani, I too am joking with you. That is not the reason. I cannot conjure these things, Ezeani, because your faith in me is narrow. It is barely enough to keep me in existence. My powers are therefore limited. I have to hustle. I am doing mail and credit card fraud. That's where the money is coming from.'

I looked at Anyanwu. I reached out and poked my fingers at his chest. He was looking at me as if slightly confused, unsure of what I was doing and what I might do next. 'You slept with Jenny! She has no idea who you are. You are defrauding people!' I screamed.

He looked bewildered. Then he shook his head. 'Ezeani, sometimes you surprise me. Is it only me?! The Greek Gods did this all the time. Even damsels much younger than these.

What about Jesus's father? Let an Igbo God start fucking the pale-skin maidens and there is immediately a problem,' he grumbled, and then abruptly turned his face away from me. Then sneered.

'Please let us talk about serious things. I want to teach you something important. You have to find a way to tell the world about me. It is crucial they understand my power; the power of our people and the Gods they worshipped. They will see that it is my acolyte, our priest-king, who walks leisurely into the Place Beyond Knowing.'

As he filled the glasses again, Anyanwu continued to speak. I drank again from my glass.

'Have you ever wondered why your title, "Ezeani", was conferred on you? Eze Ani means the king-priest of Ani. Why did I acquiesce to such a title for you? If my purpose was for you to worship *me*?' He stopped and stared at me. Then he cleared his throat.

'I have told you about the Earth Goddess Ani, who took her own life, rather than endure the slow crawl to irrelevance. She was braver than I am. I did not do what she did. Nor did I do what Ani's husband, the Sky God Igwe, the one they call Amadioha in praise, did, fighting with passion and rage, thunderbolts and storms, the slow slide to irrelevance. I, Anyanwu, I sought a way to survive. In a God, it is perhaps a shameful choice, but I am here, Ezeani, and they are not.' Anyanwu stopped speaking and stared into the distance for several moments.

Abruptly, he started again: 'The only other God in our cosmos that survived is Ekwensu, the Trickster God. Jesus and his priests knew he would. They knew because Ekwensu laughed at them, played his pranks. His mockery protected him. But they are clever – Jesus and his priests; they started telling everyone that

Ekwensu was Satan. The spirit in their cosmos who is the enemy of Jesus and his father. This is how they attacked Ekwensu. With painful libels. I don't know where he is now. His worshippers are also few. And many of them have wicked intent.'

There was a long silence. Anyanwu stared into the corner as if lost in thought.

'I sometimes wonder if Ani is gone,' he continued. 'I see signs – even here, in America – that she has come back. I know. It sounds absurd, but the resurrection of a God is not something that has not happened before. Perhaps it happened to Ani. There are things that make me wonder. Ani was fierce. She would not bend, nor move, nor accommodate this thing here, or that thing there, so that she could survive. She would not let any of the rules of the cosmos be defiled without seeking vengeance. If not on the perpetrator, then on his son, or his son's son. She wiped out whole family lines, so implacable was her rage. That is why among the Igbo the greatest sin is to defile the Earth.' Anyanwu shook his head slowly. 'I fear her. I fear her wrath if she returns.'

He poured out two more shots of schnapps. We drank. Then he poured out another round and we drank those too. He looked up at the television. It was playing short videos, each featuring the same man dressed in a suit. The chyron had changed. It now read: **C-SPAN: Senate Judicial Committee – Judge Thomas Hearings Continue**. 'They are about to start,' Anyanwu said and turned up the volume.

He pointed at the man on the screen. 'He is from our place. His ancestors are from Aro Isiokpo. They are unreliable people. Full of cunning. His ancestor was trying to sell one of his kinsmen into slavery but did not realise that he too was being tricked. This ancestor was also swallowed into the belly of the pale-skin's boat. That is how he got here.'

The television switched to a reel of a woman in a blue suit. She had an arm raised, like she was swearing an oath.

'They have accused him of trying to fuck a woman – one of our daughters. Instead of answering their questions he claims they are lynching him!'

Anyanwu stopped his glass midway to his lips and turned to me. 'Do you know what lynching is?' he asked. I nodded.

The buzz of voices from the hearing filled Anyanwu's living room. On the television screen, a man – identified by a caption that appeared on the screen as **Sen. Joseph Biden, D-Delaware; Judicial Committee Chairman** – banged a gavel on a desk and the chamber was quiet. The woman started speaking. 'His conversations were very vivid,' she said. Sitting beside Anyanwu, I watched the Senate Confirmation Hearing for a United States Supreme Court Judge and learnt humans could have sex with animals, about pornographic film and the existence of the performer Long Dong Silver. I learnt things I did not know before.

10

OUR FRIEND JESUS

A Visit from the (Thought) Police

Between the serious, regular but insular routine that characterised my life as a dedicated student, and the bursts of occasional madness inspired, and often organised, by Anyanwu, I managed in my undergraduate years at Cornell to learn physics. The things I learnt – the fundamentals of modern physics – are easy to summarise but hard to believe. At its essence are two theories:

The first, Einstein's Relativity, says space and time are facets of the same thing, and that thing is bent and folded by mass and energy – you and I, for instance. This bending means there is no definitive order in which things occur; the order of events is determined by the perspective of the observer, by its relative motion in relation to the observed. The past, the present and the future, consequently, all exist at all moments, because what is in my past in this moment may be in the future of some other being, perhaps yours, in this moment.

The second, Quantum Mechanics, holds that everything in our universe emerges from the interaction of probability waves. There are no *things*, just manifestations from these waves that circumscribe the spiritual potential of things. The mathematical description of this reality is set out in Schrödinger's wave equation. At the centre of this equation is an imaginary number. This fact greatly disturbed Schrödinger and he described its appearance in this most fundamental equation of reality as 'unpleasant'. But, as objectionable as it might have been to him or us, the imaginary number remains.

I mastered this knowledge. I learnt the equations that described these facts and the experiments that confirmed them with the exaggerated scepticism and inexhaustible hunger of one whose need to believe is deep.

It was in this state of excited perturbation that I began to experience, for the first time, long blacked-out periods that I would later hear psychiatrists refer to as a psychogenic fugue state. I could be in the postgraduate lounge in the morning, skimming through a paper, and then my time would stop. I would only become conscious again in the evening, lying still on my bed. I was not immediately disturbed by these episodes since they followed periods of intense concentration on the questions of fundamental existence that troubled me. To state this problem simply: In what sense was I, Ezeani Kobidi, real? What sense did it make that I could ask such a question?

And it was at this point in my academic progress, when the fundamentals had been established, but the question of what it all meant had only started to howl in my head, that I returned to Telluride House one clear summer evening at the end of my final undergraduate year to find two heavy-set men awaiting me in the great room. The boy with auburn hair said, 'There he is,' as I walked through the door, his pale pink finger pointing at me. They were wearing suits.

'Good evening, Mr Kobidi,' the taller of the men said. 'We are detectives from the Ithaca Police. We think you can help us on a case we are working on by answering a few questions.'

'OK,' I responded.

'It would probably be easier if we do this at the station,' the other policeman said, and started walking towards the door.

Outside on the porch we met the black-haired girl coming up the steps, two burlap bags in her hands.

'What's going on?' she asked, her eyes fixed on the lead policeman.

'Mr Kobidi is assisting us in an investigation,' he said quickly.

'He isn't in any trouble, is he?' she asked.

'No. Not at all,' he replied, with a lift in his voice.

I was relieved by the policeman's words, but the black-haired girl's face still indicated worry. I had no sense of what it meant for policemen to wish to interview me, but I felt a strong unease, wondering – not imaginatively enough it would turn out – what this could all be about.

The incidents that sprang from this experience would funda-mentally alter the trajectory of my life, lead me through so much confusion and disassociation, cause me, in a very dark moment, to question even my sanity, and commence the cascade of events that would lead me to meet and marry Heidi.

At the police station I was placed in a narrow room with a large table and four wooden chairs. The two policemen sat across from me. I was still wearing my blue polytetrafluoroethylene coat. The shorter policeman took off his suit jacket and folded it over the back of a chair. He was wearing a holster and a large black pistol.

'Do you want to take that coat off, kid? It's kind of warm in here.'

I shook my head.

'We are investigating a crime, and we are hoping you can help us identify the perpetrators. It's a case of vandalism at a church. Perhaps you heard about it?'

I shook my head.

'There's been a lot about it in the press. It's basically all anyone's been talking about for the last two weeks.'

'I don't know anything about it.'

'He doesn't know anything about all this stuff,' the taller policeman said. 'Too busy studying.' There was mockery in his tone.

'Well, you should get up to speed,' the shorter policeman said and pushed two newspapers across the table. One was called the *Ithaca Journal* and the other the *Cornell Daily Sun*. I had seen the mastheads around, on top of tables, rolled up in plastic sleeves on the sidewalk, and in a machine you fed coins, just at the door of the faculty building.

I glanced at the newspapers, uncertain what I should take from them. The shorter policeman got up, and when he stood over me, pointed to the headline article of the *Ithaca Journal*: **BISHOP ORDERS BURNING OF ALTAR DESECRATED BY PRIEST!**

The second article I found by myself in the *Cornell Daily Sun*. It described the basic facts. The Catholic parish priest of the Church of Immaculate Conception, located in the Ithaca Commons, had arranged to have a sexual orgy with a couple of transvestites atop the church altar. The act had been discovered by the janitor, Jesus Menendez, who had noticed a light in the church at 1 a.m. and had come in to investigate, expecting to find vagrants. According to Mr Menendez's account, the priest had been on his hands and knees on the altar performing oral sex on one of the transvestites while the other 'buggered' him. Mr Menendez shone his flashlight on the party, surprising them in the act and causing them to disengage. The parish priest become immediately contrite, going down on his knees and flagellating himself with a leather whip. Mr Menendez recalls that both transvestites were dressed in revealing black leather dominatrix outfits and boots. In his statement to the police, he stated that one of the transvestites ran away, but the other flew into a rage and started trashing the church, spraying the pews with graffiti. He assaulted Mr Menendez with a leather whip when the janitor attempted to restrain him. The priest, Tom O'Leary, was immediately arrested and charged with vandalism by the Ithaca Police. The district attorney described it as a holding charge, while his office considered what additional crimes the priest would be charged with. He was being held without bail at the Ithaca City Jail. One of the transvestites, the one that fled, had been identified as James McClintock, a thirty-two-year-old dentist and resident of Meredith, New Hampshire. Pressure had been mounted on the district attorney to seek his extradition from that state. The one that had assaulted Mr Menendez was still at large. The police were actively pursuing leads to locate him.

As I finished reading the story, I was uncertain what to make of it and I was wondering what any of this might have to do with me. Then the policeman slammed a photograph on the table.

'Do you know who this is?' he asked.

The grainy photograph, presumably lifted from a surveillance camera that looked out to the Commons, was the cropped image of a man in a large afro wig. He was dressed as a woman, with a black leather coat that had a large, erect collar. The photograph was unfocused, and it was hard to be certain of the person's features. But I was sure it was Anyanwu.

My conversation with the policemen lasted about two hours.

'Thanks for your help,' the taller policeman said as he opened their car's back door. 'We will let you know if we need anything else. You've been very helpful.' He smiled and shook my hand.

The next time I saw this policeman his face was hard, and

he did not smile. I was walking down Campus Avenue, off the engineering quad, when he and his partner pulled up in front of me. He got out and placed handcuffs on my wrist.

'We know you were in the church, Kobidi. You are being uncooperative. We have a material witness warrant for your arrest,' the shorter policeman said as he folded my head to avoid the roof beam of the vehicle.

'You are lucky you aren't being immediately charged with felony Second Degree Criminal Mischief. Take the opportunity and give us your accomplices. It will be easier on you,' said the taller policeman as he closed the door.

At the police station, after hours in an interview room, I was stripped of all my clothing and required to put on a white T-shirt, white underpants and orange overalls. Then I was walked down a noisy hallway and placed in a small room with a large metal gate as a door. Built into the wall was a bed. I lay down on the rough sheets and curled myself into a ball. Someone dressed in a uniform appeared at the door and shouted. I ignored him. As night fell, my stomach tightened, and I blasted at the night air with thunder rolls of flatulence.

Letter from an Ithaca Jail

Two men in green uniforms walked into the small room, then grabbed and lifted me by my armpits from the bed. I was muttering. I am uncertain of what I said. They pushed and dragged me to a room with a small table and chairs bolted to the floor. Through a small window, light poured into the room in a beautiful arc. I was looking at the light, when the men forced me into a chair. Then they placed handcuffs on me and started to chain me to the table.

'That's not necessary, Officer,' I heard a woman say, in a sweet lilting voice.

'It's regulations, ma'am,' one of the men replied. 'He has a psych hold.'

'I am his attorney,' the woman said. 'It's not necessary.'

The man shrugged and finished fastening me to the table. 'Regulations,' he said when he was done. When the two men left the room, I focused on the two people they had left behind.

My brother was across the table from me in a blue shirt with gold cufflinks and a blue suit. He was not wearing a tie. Beside him sat the woman. She was also wearing a blue suit, but her shirt was white. My brother's eyes were full of sadness, and he kept his head down, shaking it from side to side. The woman smoothed out the yellow pad in front of her with long fingers. Her face seemed familiar. I stared at her. She looked back at me. Her eyes seemed to contain a question. She scribbled on the pad as she began to speak.

'Hi, Ezeani, I'm Chijioke. Do you remember me? From Ibadan. I am a lawyer. I live in Syracuse, which is about an hour away from here.' She was still looking at me, as if she was concerned I did not understand the import of her words. My brother looked up.

'Ezeani, it's me, Nnamdi. Are you OK? Please say something?'

'They arrested me,' I responded. And then I started nodding my head. The motion continued and I could not make it stop. Memories of Chijioke came flooding through my mind. She was in a large dark wardrobe with my brother in our first house on campus in Ibadan as we played hide and seek; she was pouring out Kool-Aid in a large glass and handing it to me. Suddenly, my head stopped moving. I smiled at her. She smiled. Then she whispered in my brother's ear. I could hear some of the words – *attorney-client privilege*.

My brother stared at me. He seemed like a person that might cry. He got up and walked to me, placed his hand on my shoulder, and pulled my head into his stomach. He held me in this way for a moment before leaning and kissing the top of my head. Then Nnamdi walked to the large metal door and knocked on a glass panel. The door opened and he walked out. My head's nodding motion began again.

'How are you feeling, Ezeani? Has anyone hurt you?'

'No,' I said. When I spoke, the movement of my head stopped.

'You are being held on a material witness warrant. They haven't charged you with any crime and that is a good thing,' Chijioke explained.

She looked down at a manila file beside the yellow notepad and opened it. Then she continued speaking. 'The police say a witness saw you loitering outside the Immaculate Conception church on the night it was vandalised. The witness, a Jesus Menendez, is clear in his statement that he never saw you enter the church. However, you did not mention this to the police when they interviewed you. Given the inconsistency of the information you provided to them about one . . .' – she flipped a page in the file and then read – 'Anyanwu – AKA Sun – they think you are involved and know more than you have told them.'

I did not speak.

'Frankly, it's an abuse of process. Unfortunately, it happens quite often,' she continued. 'They should never have been able to get that warrant. The community is really agitated, and the police and prosecutors are under a lot of pressure so they are cutting corners.' Her brow furrowed but her voice did not falter in tone.

'The police are now getting lots of flak for arresting you. They claim they have not been able to interview you because of your erratic conduct. They are concerned about your mental health and would like to do a mental evaluation. I think everyone now realises what a mess this is.'

I was fascinated by the sound of her voice. I could see in her face the Chijioke I knew as a child. It was in her large eyes and the slope of her nose. And the eyes and nose connected me with a child I had also once been, and this comforted and soothed me. But the voice was new. It sounded sweet and rising, but with an undergrowth's murmur. It did not connect me with the child I had been or a Chijioke I had known. I had no memory of Chijioke's voice. Time passed.

'How long have I been here?' I eventually asked.

'Two nights. Nnamdi found out yesterday evening and called me. I drove down this morning. I have brought a motion for your release before a judge this afternoon. I hope it happens before 4 p.m. so you can be processed and don't have to spend another night here.'

I did not speak. I smiled at her. On the yellow pad, behind her left forearm, she had written: **Dilated pupils? Schizophrenia? Mania?**

'Let me go and see about the hearing. They will keep you here till they bring you down to court. It's good to see you again, Ezeani. Not great circumstances, but it's good to see you.' And then she rose and walked to the door. The heels of her black shoes clicked against the grey floor in a hollow, round sound. She turned to me and smiled: 'Don't tell Nnamdi, but you were always the handsome brother.' And she waved at me, her face still fixed in a wistful smile. The heavy door opened and Chijioke walked out of the room.

I sat on the chair looking at the sunlight pouring in through the small window. This light had left the sun approximately eight minutes ago and was now flowing into this grey room, casting a partial rainbow on the floor. I stopped myself. This was just poetry. It was important that I did not lose sight of the fundamental facts: permutations generated by the sun in the electromagnetic field had travelled as a wave rippling out in all directions. When the electromagnetic wave interacted with the quantum waves that described the event called 'Ezeani Kobidi in an Ithaca Jail' they generated this magic entangled event – 'Ezeani Kobidi in an Ithaca Jail Seeing Light'. My mind immediately went to something I had read in my father's study, *Letter from Birmingham Jail*. 'I am here because injustice is here,' I said to the empty room.

I am here because quantum waves are here! I had an existence, there is no other word for it, because the photons of light and the electrons of the chair interacted with the probability waves that in their quantum state reflected everything I could possibly be. This interaction was what it meant that I existed.

I was pleased with my thoughts. With the brutal, unsentimental honesty with which, it seemed to me, I was able to analyse the situation. These were the only reasons I existed in this moment. Different, random combinations would mean a different existence. If you ran the probabilities for long enough, in one of them, I would just puff and disappear.

If you learn physics, you will know all this is true, but you will object that the probability I will disappear is so infinitely small that it might as well be ignored. And my response to you will be: Why should it be ignored? Doesn't it tell you everything that you need to know that such a possibility actually exists?

The room was silent. I busied myself calculating precisely the probability that I would just disappear. Occasionally, a screaming human voice would rise sharply for just a moment and then die down again, leaving quiet behind. This went on for a long time. I closed my eyes.

When I opened them, seated in front of me was Anyanwu. I screamed. A guard's face appeared momentarily in the door's glass panel, peering briefly into the room and then moving away. Anyanwu was wearing a blue suit, a blue shirt and an elegant tie. He looked very neat, his beard cropped and lined. On his head was a powdered lawyers' wig, the type worn by English and Nigerian barristers. In his right hand was a large lawyer's briefcase.

'Why are you screaming?' Anyanwu asked me, as he leaned into the table. 'Please keep your voice down. I am not going to hurt you. Did you tell them anything?'

I stopped shouting and stared at him.

'You have the right to remain silent. They can't make you talk. The whole thing makes me so angry.' Anyanwu placed the large briefcase on the floor beside his chair.

'Look at the way they have tied you up. Like a slave!' he spat. 'Don't worry. We will make them pay! We are going to sue the hell out of them! They will be paying millions for this. On the good side, I am sure Jesus is not finding any of this funny,' Anyanwu laughed. 'I was selling pills to that his priest. That's how I saw the opportunity! He is not finding it funny at all. "Immaculate Conception"! that's my new nickname for him. When I saw him the other day near Ithaca Commons, I shouted, "Immaculate Conception"; he turned and threw the drink he had in his hand at me. It was so funny.' He laughed louder and slapped his thigh. I did not respond.

Anyanwu stopped talking. He was looking at me closely. 'Are you OK?' he asked. His face was now serious and concerned.

'I don't know what is happening to me,' I replied. 'I am afraid.'

Anyanwu looked at me and nodded knowingly. And then he smiled. 'I am happy that you have told me what is troubling you. You have to be careful, Ezeani. When one sees nine mad people walking down a crooked path and decides to follow them, one should know that a passer-by will report that they saw ten mad people walking down the crooked path. It is important that you do not succumb to the traps that swallow the unwary.'

He reached into the briefcase and brought out a large palm wine gourd, laid it on the table and took out two drinking horns. He blew into the empty horns as if he wanted to expel dust. Then he filled one of the horns with palm wine, swilled it and poured out a libation. The white palm wine foamed on the concrete floor. I saw small rainbows in the bubbles. 'Drink, Gods and Ancestors! May it be well for you, and may it be well for us,' Anyanwu said.

Then he poured wine into the drinking horns and passed one to me. The aroma filled my nose. I drank the horn and handed it back. 'You are thirsty,' Anyanwu said, and smiled. He poured out the frothy palm wine till the horn overflowed and wine touched his fingers, and then he gave it to me. I drank.

'Let me tell you a story that will clear your mind. It is an account of things that happened. Perhaps if you understand the story, you will understand where you are. In the time when the universe was in the phase before this one, when the universe consisted of one thing, the singularity Oma, the Oma's Chi, its spirit and its Eke, the Oma's manifestation, started to separate and become distinct. The Chi grew within itself, ruminating on thoughts that were the Chi's alone. Chi did not share them with Eke. Consumed, Chi started building an Upright Lodge to shelter his thoughts. Chi entered the Upright Lodge with his back and emerged from it with his back, so nothing on his face could betray what he saw. In the Upright Lodge, Chi's thoughts could gyrate and contort, unobserved by the universe.

Chi forbade Eke from entering the Upright Lodge. "What wanders about there belongs only to me," he declared.'

Anyanwu stopped. 'I am sure I do not need to remind you of the nature of the Parent Godhead, Chi ne Eke?' He looked at me. I said nothing.

'One day when Chi was away, Eke, consumed with curiosity and jealousy, opened the door of the Upright Lodge. A massive explosion, beyond the force of a thousand, thousand suns, rent the fabric of Oma. Chi came back to find Eke dead, her body scattered in a million pieces around the Upright Lodge. Chi was instantly overcome with sorrow and guilt. He walked into the Upright Lodge and closed the door. He has never emerged. The door of the Upright Lodge has remained shut ever since.' Anyanwu lifted sad eyes to look at me.

'But something else happened: the dead Eke – the Goddess of creation and manifestation – began to come back to life, to manifest into new things that kept growing and filling the universe. Each of these things had the attributes of the Parent Gods, Chi and Eke. Each had manifestation from Eke and each had spirit from Chi. In the Upright Lodge, Chi felt a stir the moment that his beloved Eke resurrected. Through the walls of the Upright Lodge his spirit connected with her, and a part of Chi was connected to everything Eke manifested. But despite the deep and abiding intimacy of this connection, Chi kept the door of the Upright Lodge firmly shut. So great was his fear that he would again destroy his beloved Eke – and with her, all of creation which had manifested from her. Chi and Eke, spirit and manifestation, filled the universe. And it remains this way to this day. That is why in this phase of the universe everything comes in twos, each with its Chi – its spirit and potential – and its Eke – its manifestation and realisation.'

I handed Anyanwu my empty horn and he filled it with palm

wine. I had barely lifted the horn to my lips when I passed out. My time stopped.

The Saviour of the World

The judge sat on an elevated platform behind a high bench. I was seated in a wooden chair beside Chijioke. Sometimes when Chijioke spoke, the judge's ears reddened. Her voice was clear and firm. She hardly moved her body, her long fingers barely touching the yellow pad.

Behind me, in the first row of what looked like pews, was my brother Nnamdi, the woman with red hair and Rayburn, the professor with grey hair.

At various times, Chijioke asked all the people in the front pew to stand and speak. She would say some things about me to the judge, nice and complimentary things, and then she would ask the judge's permission for one of the others to speak. They would get up and repeat, in different words, and perhaps not as concisely, the same things Chijioke had said.

At another desk almost beside us, a woman and a man in dark grey suits represented the district attorney. They did not seem opposed to Chijioke's arguments, but they would stop the proceedings regularly to make some clarifications or requests whose meaning and import was entirely lost to me.

Then I heard Chijioke draw a long, strained breath. I looked at the judge. His ears were a fierce red, with blotches of pink. He drew himself up on the bench. He lifted his right arm and pointed it in my direction and then moved it up and down as if he wanted me to levitate.

Chijioke leaned down and her breath was in my ear. I

could smell her sweet perfume. Her voice was low. 'The judge wants to ask you some questions,' she said. 'You have to stand up.'

I pushed the muscles of my legs to stand. When I accomplished this, I noticed that the judge had been speaking. I focused on his mouth and soon I could understand what he was saying.

'In fact, I think I should make it plain that I worship at Immaculate Conception. I, and every member of the congregation, am deeply distressed and shocked by the criminal and, frankly, abhorrent activities that took place there. I understand that you are a very promising undergrad expected to return to Cornell for a PhD. However, I am still not satisfied that you have been entirely forthright on what you know about these events.' He stopped speaking and looked at me inquisitively. 'Do you understand what I am saying, Mr Kobidi?'

'I believe I do,' I responded.

'Good. A lot of people have vouched for your character. There has been some suggestion that you might have some challenges in communication and socialisation from perhaps a mental health perspective. I am very mindful of that, but I must insist we get all the information that a society has the right to request of its citizens. Are you following me, Mr Kobidi?'

'I believe I am,' I responded.

'Good. Very good, Mr Kobidi,' the judge said, starting to smile. 'We are making progress.' I looked at Chijioke and her face seemed stern. She was tapping her long fingers gently against the yellow pad.

'What do you know about the vandalism at the Church of Immaculate Conception on the night of July 28?'

'Do you believe that Jesus is the Saviour of the World?' I asked the judge.

The judge was quiet for a moment. He seemed surprised. 'Yes, in fact I do, Mr Kobidi,' he said.

'Can you please explain to me how that is true? I would like to understand.'

'Mr Kobidi, I would love to do that, and perhaps I will get an opportunity to do so under different circumstances. However, as a judge sitting in on this hearing, and bearing in mind the constitution's separation of church and state, I am required to go back to my earlier question. What do you know about the events of July 28?'

'Do you know why the world needs to be saved?' I asked.

Chijioke's voice rose slowly. 'Your honour, may we approach?' Then she and the lawyers from the district attorney's table walked over to the judge's bench. They whispered together for a long while. The judge's ears reddened as they spoke. Then the lawyers walked back to their chairs.

The female lawyer at the district attorney's table did not sit down like Chijioke and the other lawyer. After a few moments passed, she said: 'For the record. We confirm that Mr Kobidi, represented in this proceeding by Chijioke Yemisi, has been granted immunity by the Tompkins County District Attorney to testify truthfully to all facts he may possess in connection with the events of July 28 at the Church of Immaculate Conception. The court has requested a psych evaluation of Mr Kobidi to ascertain his competence to serve as a witness, after which he will sit for an interview with the district attorney's office to discuss the facts of this matter. The interview shall take place on or before September 28, unless it is the recommendation of the psych evaluation that the interview be postponed to a different date. The district attorney's office shall not commence any criminal proceedings against Mr Kobidi as long as he is in compliance with these conditions. In light of the foregoing, we have no objection to Mr Kobidi's release on his own recognisance.'

Immediately the woman finished speaking, the judge banged his gavel. 'ROR,' he yelled. Then he turned his face to me. 'It has been a pleasure speaking with you, Mr Kobidi. You ask interesting questions. I wish you luck. Good day.' He shot up from his chair and quickly left the room.

Chijioke leaned down. 'The judge just ordered your release. It's past 4 p.m. now. I just hope you don't have to spend another night in here.' She seemed worried and hurt. I smiled at her.

Anyanwu the Seductress

It would turn out that the judge's order had come too late, and I would spend another night in the Ithaca City Jail. I was held in court for many hours while Chijioke moved from room to room in the building. My brother took her vacated seat, placing a large paper bag on the table. He reached into it and brought out two large paper cups filled with orange soda, and hamburgers and fries enclosed in cardboard boxes.

'Is that Fanta?' I asked.

'Your favourite,' he said and smiled. 'No, it's Mirinda. But it tastes like Fanta.' We ate and drank.

As we ate, he spoke. I do not recall all the words, but he said my sister had been unable to come. The reason was serious and clear to him, but I did not really comprehend it.

When officers in green uniforms arrived to take me back to the Ithaca City Jail, my brother grabbed me and hugged me. He was crying. A few tears. Chijioke stood behind him. She looked sternly at the men. Her voice was measured and harsh when she spoke. They held me by each arm, raised me up, handcuffed me and then walked me through a long corridor to a glass door. The sun was large and orange as it dipped its ends below the horizon.

I had barely fallen asleep on the bed clamped to the wall when I felt rough slaps on my heels. My sleep had been greatly disturbed by the clangs of the closing metal doors, the howls and other human cries that came in from the long corridor. I resisted consciousness. But the slaps persisted. I opened my eyes. It was Anyanwu.

'You have woken? Good. You did very well today. You didn't tell them anything.' He laughed and then slapped his thighs. 'Instead, you asked them questions!' He had undone the collar button of his blue shirt and his tie hung loosened around his neck. The white, red and black floppy cap had replaced the lawyer's wig.

'I know you want to sleep, but I just wanted to tell you some things so no one can trick you. They have your fingerprints inside the church. Jesus will swear that you did not enter the church, though. You don't have to worry about that. I threatened him. He will keep his mouth shut. The rest is up to you. Don't fall for their tricks. Don't tell them anything.' He took a bite from a lobe of kola nut and started to chew.

'I can't tell them anything! I don't know what happened!' I shouted in irritation.

Anyanwu stopped and considered me. 'You really don't remember what happened?' he asked, a quizzical look on his face.

'I don't know,' I responded.

Before I could stop him, he placed his rough palms on my shoulders and blew a froth of masticated kola nut and saliva on my face. The jail walls fell away and I was covered in darkness. Then I heard the low hoot of an owl and felt a night breeze on my cheek. I opened my eyes. Across the Commons was the Church of Immaculate Conception. Anyanwu was seated on the bench beside me dressed in black leather. He had red lipstick on his mouth and he was wearing a large woolly afro wig.

'Don't be so surprised,' he said to me. 'It is not now that I learnt to dress as a woman. Among our people who dwell on the far side of the Urashi River and those that dwell south of the lower parts of the Imo River, they know me as a woman. And it must be so, for I appear at their festivals as a woman. As the Parent Gods of our cosmos are paired, husband and wife, Chi and Eke, so are my fellow principal Gods. Ani, the Empress Goddess of the Earth, has her husband Igwe, the Sky God that is saluted as Amadioha and Kalu. But I, Anyanwu, am paired with no one. I, the God of Sight, the All Knowing One, even I, do not know why this is so. So the pairing that Chi and Eke did not create in one way, they created in another. When I go down to the festivals of the low-lying lands, I manifest a woman, wear bangles, and go out and greet them.'

Anyanwu's eyes seemed to be waiting for something from me. I was irritated. The church was barely lit by the lights from the street.

'What are we doing here?' I asked.

'I am waiting for the other man to come,' he said, and opened his right hand. A few pills lay in his palm. 'Jesus's priest swallows these like they are food.' He chuckled, his mood starting to lift. 'They loosen him up, like alligator pepper when it has entered your brain. That is how I found out what the priest wants to do in Jesus's shrine. If I tell you, you will not believe me. That is why I have brought you here. So you can see with your own eyes.' He started laughing, so loud that it frightened the owls to silence.

*

The next day, for obscure reasons connected with missed or incorrect paperwork, I would be threatened with the possibility of another night in the city jail. Through the morning and a great portion of the rest of the day, Chijioke would appear in the interview room to reassure me she was doing everything to secure my release. She informed me that my brother, who had booked a room at the Statler for use when I got out, had been obliged to return to New York for work. He would be back over the weekend. I was surprised to learn that Chijioke had not spent the night in Ithaca but had driven back to her home in Syracuse the night before.

'It's only an hour's drive and I don't like to be away from my family if I can help it,' she said. 'Nnamdi booked a connected suite. It's a really fancy hotel. I will stay with you at the Statler until he comes in tomorrow.'

11

IN ORDINARY TIME

The Second Sunday in Ordinary Time

Time is only felt by things that have mass; it runs at different speeds in different places; it runs at different speeds for those moving in different directions; it ticks at different rates for those travelling at different speeds. Time forces us to fall when we stand near massive things. Time slows down for some. Time stops for others. These were the things that were known about time before I published my third brilliant paper.

This article cemented my reputation as one of the leading theoretical physicists of my generation. I published it under the provocative title 'The Length of Time'. In this brief but seminal work, I established through exquisite reasoning and clear mathematics that time is an emergent property of our universe. Like all emergent properties, it is subject to interpretation. Time is like the notations in the notebook of an accountant at a casino run by mobsters. It records something – but what it records is an impression from a particular, not well-informed, point of view.

The three papers would create an air of promise and anticipation about me. Along with my thesis, they would assure my position on the faculty at Cornell University. I did not then understand that the work was incomplete. I did not realise that the deceptive sense I had as an adult of the flow of time, very different from the choppy improvisation of my childhood, had misled me.

I sat quietly in the wooden pew. Heidi, the woman who would become my wife, sat beside me. The words of the priest washed over me in a soft, melodious tenor.

'This we ask through Jesus Christ, our Lord, who lives and reigns with you, in the unity of the Holy Spirit, one God, now and forever,' he said.

As the priest finished speaking, Heidi and the twenty-one other people in the church, almost in unison, said: 'Amen.' Heidi reached out and held my hand.

We had lunch on the sidewalk, outside a restaurant with blue awnings. It was late summer and the air was cool, the wooden tables were covered with white tablecloths, and the waiter wore a bow tie. Heidi ordered the meals for both of us. She held my hand as she spoke to the waiter. When the bowls of pasta were placed before us, I unwrapped the tableware from a large white napkin and started to eat in silence.

Those were the early days, when Heidi nurtured and protected me, fought fiercely to establish my reputation in the field, convinced that my genius was being diminished by factors and forces around us that she claimed to understand far better than I did. This was before the birth of our son, our one and only child. Before she would – in exasperation, fear, ambition, vulnerability, I never understood her motivation – inexplicably turn against me with a ferocious single-mindedness that I must admit entirely overwhelmed me.

Though I learnt physics and developed several deep insights that have moved the science forward, I must admit that the naivety and confusion of my childhood remained with me. I was incapable of assessing the forces that led those around me in the confusing, varying velocities in which their lives travelled, pushing and pulling me along, like planets caught in the same gravitational well.

Isaac Newton, when he pondered his equations of motion, which neatly modelled the effects of gravity on our solar system, was perplexed by one thing: What invisible force explains the instantaneous pull and push of the sun and planets on each other, resulting in this, and I quote him directly, 'action at a distance'?

Albert Einstein's genius insight was to see that it was the space and time between the sun and planets *itself* that created this invisible pull. The significance of Einstein's theories of relativity is simply this: there is no *where* without a *when*. They are coördinates in the same thing, the same field, the field we now call spacetime. And this spacetime bends and folds against the mass of heavy things, drawing them to each other. As you know, when Einstein was confronted with the peculiar effects of quantum entanglement that resulted in unknown forces from across the universe being able to affect, instantaneously, activity on Earth, he described it, and here also I quote directly, as 'spooky action at a distance'.

The genius of my crowning achievement in physics, the work that would eventually bring derision and calumny to my reputation, was to see that it was this quantum entanglement *itself* which created space and time. You see, my mother, father, sister and brother share the same space and time with me because we are entangled; I share space and time with my wife because we are entangled; and dear reader, as you are beginning to see, even though I am dead, perhaps long dead, you and I share space and time because we are also entangled.

It is important that you understand what I mean. We are *entangled*, and the manifestation of that entanglement created the space and time, the *where* and *when*, in which you started to read this book. When you see this clearly, you will begin to understand what my final equation means.

What we know is eclipsed by what we feel.

The Blessing of a Brother

The bed was enormous. I rolled myself from one end to the other, luxuriating in the clean, fresh-smelling sheets. The contrast between this large bed and the narrow metal frame fastened to the wall of the Ithaca City Jail felt like a phase transformation of some sort. The same substance, but reformed in a new, intriguing way, like water that had become ice.

The first night, I slept alone. Chijioke was in an adjoining room connected by a thin door. 'I am just on the other side. If you need anything or if you are afraid, you can call me,' she said, as she slowly closed the door.

I threw myself on the bed and, almost immediately, fell asleep. I was awoken at 2.11 a.m. by a persistent knock. When I opened my eyes and glanced at the digital clock on the nightstand, I was disoriented. I rose from the bed, walked to the front door in the suite's living room and looked through the peephole. In the curved convex image was a waiter in a formal white shirt and black bow tie, standing, his heels rising and falling impatiently.

'Yes?' I said.

'Room service,' he responded.

I opened the door and the waiter rolled in a trolley. He lifted a large tray with many dishes and two bottles of Mirinda, and set it on the coffee table near the window in the living room section of the large suite. 'You must be hungry. It's a lot of food this late at night,' he said and smiled. Then he presented me with a small black leather folder. I hadn't called down to order any food. I was confused. It was unclear to me what was expected.

'You need to sign,' the waiter said as he flipped open the folder and handed me a pen. I signed.

'Just call the desk when you're done. We'll come up to clear the tray,' he said.

I walked the waiter to the door. As soon as I shut it behind him, I heard a sharp popping sound and then the hiss of one of the soda bottles being opened. I turned around. Anyanwu was lowering the bottle he had just opened with his teeth to the table. He was no longer wearing a suit. He was dressed in a yellow tracksuit with a white T-shirt visible underneath the jacket. Around his neck was a large sun medallion, dangling on links of wrought gold. On his feet were white tennis shoes and, on his head, a white knit skull cap through which was inserted a large eagle feather. 'This is a nice room,' he said. 'Let us eat.'

We sat and ate together. We did not speak. When we had finished, Anyanwu cleared his throat and said: 'They have finished carving your Ikenga! The time for play is over. You have to gird yourself. This is the time for you to be stout-hearted. That is what I have come to tell you. Fortify yourself so the world will hail you! And when they hail you, they will start to remember me, your God!'

I took a large drink from my soda bottle. Then I belched.

'Ezeani, when did you last hail your sister Obiageli?' he asked. 'I notice that your brother has come many times to visit you. Your sister is not here. I have not set eyes on her in America.' I just sat and stared at him.

'Ezeani, I asked you a question. You are a titled man, and you are no longer a child. You should answer. With your high title, you became an Nze. This means that you must keep away from acts of corruption. You should avoid the company of the corrupt and keep virtue by your side. The deep, uncompromising virtue of Ani, Empress Goddess of the Earth. I have told you this before; perhaps, she is still here with us. This is why I ask you: "When did you last hail your sister?!" Your hands must be clean before Ani. You should not be part of any abomination on the Earth.'

I was entirely confused and could make no sense of the words from Anyanwu's mouth. He kept speaking at length in this manner, alluding to the need for justice and cleanliness. His chief concern seemed to be that nothing interfere with the path of my success so that I could acclaim him to the world. 'I know things will not go back to the way they were before. But when they understand what you have accomplished, I will have many worshippers.' He was speaking with a frenzied earnestness.

Then, tiring of his words, I said: 'It seems to me you are the most corrupt person I know.'

Anyanwu stopped speaking and fixed his eyes on me. They were wide and seemed to be full of rage. Slowly, a smile developed on his lips. Suddenly, he started laughing. The laughter grew louder. Abruptly he stopped.

'Ezeani, my acolyte, it is not so. Among the spirits you know, humans or Gods, I am not the most corrupt. It is true I have done this or that other thing so I can keep surviving; I have played this or that other game with Jesus and his priest, but these are little games against the horror that Jesus and his priests have inflicted on our people.'

Anyanwu stood up, pulling down the zipper of his tracksuit and lifting his shoulders. He cleared his throat. 'There is a certain prophecy that we know. It was bandied about us when they started carrying off our people in wooden ships; when each hamlet gathered around itself fearing both the friend and the stranger. It promised our people a saviour. I did not take it seriously. I stopped this prophecy from spreading among our priests. Ani would berate me. She would tell me that my arrogance would end up destroying us all. But I did not let them share because I did not believe. Now, I know I was wrong. Even if I did not believe, I should have shared. For prophecies do not have to come true to change the world. Powerful movements are born from prophecies that don't come true; look at the religion of the followers of Jesus and the unfulfilled prophecy of his return to save the world.'

'You mean "yet unfulfilled" prophecy,' I said.

Anyanwu laughed. 'Ezeani! I hail you! Your tongue is sharp! You and I know he isn't coming back to save the world. No one but crazy people believe that any more. He is wandering around the place like me. His powers are small. The real believers are few. That is why he even bothers to talk to me.' Anyanwu stopped and looked down at the carpeted floor. 'Perhaps he wants to know how it feels; how a God can watch its people look down on their own God,' he spat, bitterly.

His face scowled for many moments and he seemed to be considering the implications of what he had just said. He looked at me. Then his eyes darted from one corner of the room to the other. He stood up and started pacing, beating on his chest in a steady, heavy rhythm. Abruptly, Anyanwu turned and ran towards me, grabbing me by the shoulders. He started to shake me violently. 'Ezeani! What if Jesus is after you?!' His eyes were wild, and spittle had formed on the edges of his lips. 'What if he wants you too?! Perhaps that is what he is after! He and his priests are full of cunning. Maybe he is trying to see if you will save him; you and your friends in physics. Don't believe him and his lies. They are deceivers.' He continued to shake me and then blew spit in a fine spray on my face.

I travelled through many visions. I was on board the wooden slave ship and lashes stung my back; I was a child hunting a small grasscutter in Umudim when a net was thrown over me; I held my bleeding arm, staring at the stump where my missing hand would have been; then I saw others like me; they were like balls of fire, disembodied things; I knew they were people because of their voices, thousands of voices crying out in pain; the sound kept resonating, growing louder in successive waves. Gradually, each voice became distinct; the piercing sounds filled my consciousness; they were millions and millions of voices, and yet, there were more; then, before my eyes, the balls of fire took shape, growing larger and taking human form: men and women, all naked, millions and millions of them arranged in rows, beyond the point my eyes could see, their skin glistening with sweat in the moonlight. The terrible wailing continued, and the sound started to form the same notes and then the same words. They all raised their arms, men and women, and pointed. I could hear the words now, in a deafening chorus: 'The Saviour of the World.' I turned to see. Before a large plume of grey clouds, floating on a stage was a large cross. The pale man walked towards the cross and stretched out his arms, as if warming his limbs. His assistants, also pale-skinned, helped lift him up and drove nails into his arms and feet and the red blood spurted. He was smiling, like a person receiving applause. Behind me I could hear the wails continue. Now, the words were mumbled in many different tones, in a multitude of languages. They were shrieking in pain, their hands still pointing. They were many hues of brown and bronze and copper and yellow, all glistening with sweat. I turned again to the man on the cross. He was looking vacantly at the multitudes. Then he saw me. He started laughing. The laughter was in his mouth and his eyes. 'Don't listen to

Anyanwu. He is just jealous. Let's talk,' he said to me. Then he winked, his mouth twisted in a grotesque smile. It was at that moment that I passed out.

I would not awaken until late morning.

When I woke, my brother Nnamdi was seated in a chair beside the coffee table in the living-room section of the suite. He had a coffee cup beside him and some papers in his hand.

'How did you get in?' I asked.

My brother smiled. 'Ezeani, the suite is in my name. I got a key at the desk.'

I stared at him. 'When did you get here?'

'Probably just before 5.45 a.m. I drove straight from New York. No one on the roads.'

He gestured at a large tray on the floor, full of soiled plates, many with half-eaten meals. 'I see you have been making good use of room service,' he said, and smiled.

We met Chijioke at the front desk checking out of her room. My brother walked up to the clerk and presented his credit card. 'I can just bill you for this,' Chijioke said.

'You and I know it's unlikely I will ever see a bill from you,' my brother responded.

Chijioke laughed. 'Are you checking out, too?'

'Nope. We are driving to Skaneateles Lake. I am going to get a boat. Ezeani and I are going fishing.'

'Well, in that case, you can drop me off in Syracuse. My husband can't make it back till mid-afternoon.'

As we pulled up behind a grey minivan in the driveway of Chijioke's house in Syracuse, her husband and their two children came out on the porch. Nnamdi helped lift her small black overnight case out of the car and walked up to Chijioke's husband

and shook his hand, hugged Chijioke and high-fived the children. They waved at us as we drove off.

'Our adventure begins!' My brother slapped his hand against my knee. I looked at him and I smiled.

There were just the two of us in the large boat. Once we cleared the marina, my brother asked me to take the wheel and steer the boat out to the further reaches of the lake. I stood on the raised helm, looking at the still water and the horizon that lay above it. I felt the slow lake wind against my face. When we were between inlets, Nnamdi stopped the engine and dropped anchor. We placed fishing lines in the water, ate our sandwiches, and talked. We talked about Chijioke, talked about our life in Ibadan; we talked about our dead mother, our distant sister and our living father. My brother told me many things. The most important was this: 'Don't let them defeat you! Your struggle is special.' He also told me not to be afraid of the psychiatrist the judge had ordered me to see. 'Listen to him. If what he says is useful then use it. If it is not, ignore him.'

The tranquil lake barely moved, and we were all alone. 'I love you, Ezeani,' my brother said. Then he started to cry. We stayed out on the water, casting our fishing line, and watched the sun set. Surprisingly, I caught a fish, my brother helping pull in the taut line.

At a restaurant overlooking the lake, my brother insisted the fish, wrapped in the discarded sandwich paper, be prepared and served with our dinner. The waiter told my brother it would not be possible. He cited many reasons, and I heard something about food regulations. But my brother would not be deflected. The owner, summoned at Nnamdi's insistence, reluctantly agreed after they negotiated a surcharge. The meal of steak, lobster and our little fish with large sides of French fries was delicious. We drank a bottle of wine.

When we arrived at the hotel in Ithaca it was almost midnight. My brother and I crawled into the large bed. I slept till morning. Anyanwu did not come.

The next morning, Nnamdi and I had breakfast in the hotel's large dining room. The windows looked out to a sloping lawn. As he sipped his coffee, my brother's hands played with a brochure. 'There is a wonderful bird reserve close by. I think we should see it.'

I continued sipping my orange juice.

'You need to go out in nature more,' Nnamdi said. 'Don't let your thoughts be squished in these small rooms.' Then he told me about the large bonus he had made trading credit derivatives.

The Couch

The red-haired woman walked me to my first meeting with my psychiatrist. The plate on his door read: *Dr Chan, PhD.* She waited outside the door of the small office until I walked out about fifty minutes later. As I zipped up my coat, she asked, 'How did it go?'

I smiled and said, 'Fine! He seems very curious about my mother.'

I would visit Dr Chan once a week in the following years. The first year was a requirement of an arrangement Chijioke had negotiated with the judge and the prosecutors. We would return every three months to discuss what everyone in the courtroom euphemistically referred to as my 'progress'.

Chijioke would drive down from Syracuse for these brief

conferences. They were all the same. The judge would read Dr Chan's report and deem it satisfactory and then ask for comments. There were none.

One day, I waited in the courtroom beside the red-haired woman as the judge went through case after case. Chijioke was late. When she finally appeared, the judge said, 'Let's do this in chambers,' rising from his elevated desk.

In the small room, the judge took off his black robe and hung it on a tall coat stand, then settled himself in a large burgundy armchair. I sat down on the only other chair in the room and Chijioke stood behind me.

The conference went like all the others. At the end the judge suggested Chijioke no longer drive in. 'We can do it in chambers,' he said. 'You can just call in from Syracuse. Good seeing you again, Ezeani. I'm really impressed with the progress you are making.'

Chijioke patted my shoulder and indicated I should rise. As we walked towards the door a pretty, blonde-haired woman opened it. 'Hey, Dad,' she said, and then stopped. 'I am so sorry. I didn't know you had people in here.' She smiled apologetically at Chijioke.

'Oh, it's OK,' the judge said. 'We are done. Counsel, this is my daughter.'

This was the moment I first saw the face of the woman who would become my wife.

A Visit from the Distinguished Professor

The amplitude of a wave is the distance between its middle position and the most extreme position in each direction. The amplitude of a quantum wave is its possibility. The square of that amplitude, i.e. the square of the possibility, is the *probability* of its manifestation at that location.

It was my brother Nnamdi, in the second year of my PhD, who informed me of my father's visit. This was after I had published two of my three brilliant papers and had attained a reputation in my field that could easily be assessed as bright. I had not spoken to my father since I left Ibadan. Often, in the intervening years, Nnamdi would pass on his voluble complaints: 'Tell him to call me! He makes no effort to stay in touch.'

Nnamdi told me, with sympathy in his voice, 'I really feel sorry for him. He said it was difficult when you were going through your episodes in Ibadan. He didn't know how to help you.'

Nnamdi was paying our father's expenses in the United States. He arranged for me to meet him in a coffee shop attached to the Statler Hotel. My father was wearing a grey jacket over a blue vest and tan trousers. He had shaved his beard and had different glasses. This made him appear strange to me. He held out his arms to embrace me. I walked past him and sat down on the wicker chair on the other side of the small table. My father turned and sat across from me.

'I am so proud of you, Ezeani! Everyone speaks highly of your work.' He paused and looked around the room. 'You know, you are the only one of my children that took after me. Academically, I mean. Your academic focus is gratifying. Gratifying.' Then he motioned to the waitress. After a brief conversation with her, he ordered a coffee for himself and an orange juice for me.

'This place is so expensive,' my father said. 'I don't know why Nnamdi insists on wasting his money like this. If he thinks it impresses me, he is mistaken. He should be using his talents for something that's actually useful to society, instead of gambling in a casino.' Then he looked down at the menu. 'I think I will get the chocolate cake. Do you want something to eat? The lemon tart cake is very good. I tried it this morning.'

'No,' I responded.

'Do you think you will stay at Cornell after your PhD? I spoke to some of my contacts, and I understand that the best physics is being done on the West Coast.'

I didn't respond.

'You should think about it,' he said.

We sat for a few minutes in silence. The sound of clinking plates and cutlery filled my ears. My father stared at me. A small smile twisted the corner of his lips. 'I have something for you,' he said, as if he had just remembered. He reached into a leather bag beside him. 'It's a gift from Umudim. Three of the village's titled men delivered it to me a few years ago.' He placed the object, wrapped in checkered red, white and black cloth, on the table.

I reached out and pulled the object to me. My father laughed. 'How do you communicate with your professors and colleagues if the only words you speak are "Yes" and "No"? I suppose since language is my area, I find it hard to comprehend.'

The waitress stopped at our table and placed a large coffee in front of my father and a glass of orange juice before me. Then she reached into her apron, lifted a straw and laid it beside the juice.

As she walked away my father asked: 'Have you spoken to your sister? I have been calling her. She has not responded to even one of my calls. So hard-hearted. If there is anything I have done to her, she should forgive me. I am not an ogre. I am getting old.' His small eyes glistened behind his glasses with moisture as he told me the story of his own suffering at the hands of Obiageli, my sister and his daughter.

I turned away from my father and lifted the object from Umudim. I started to rise.

'Please, sit down, Ezeani,' my father said.

'No!' I responded.

'Please, sit down. I haven't spoken to you in years, my son. I know things are difficult, but the distance that has developed between us is heartbreaking. Has Obiageli been telling you things?'

I started to move.

'Please, Ezeani,' he said and reached for my hand. 'I don't know what your sister has told you. I want to be upfront with you.'

I pulled my hand away. Then I zipped up my polytetrafluoro-ethylene coat.

'I was drunk,' I heard him say as I passed the table, walked through the door at the Statler and back to my room at Telluride. The next morning, I unwrapped the package from Umudim. It was a short wooden carving. My Ikenga.

The Teaching of Jacob, the Newly Baptised

My brother and I checked out of the Statler the day after our outing on Skaneateles Lake. We spent the morning at the Sapsucker Woods Bird Sanctuary and then drove down to New York City. I would spend the summer between my final under-graduate year and the start of graduate school with him in New York. Outside of childhood, those two months would be the longest period of my life I would spend in the continuous company of another human being. I say this without excepting my marriage. There is perhaps something to note in the fact that this period, as was the period of constant companionship in childhood, was spent with the same individual – my older brother, Nnamdi. He took a leave of absence from his firm, and we spent each day, from the moment we rose to the moment we went to sleep, together.

There was one notable exception. One morning, my brother and I left mid-morning for the Metropolitan Museum. It was our habit that summer to go through the cultural and historic sites and venues of New York City. We took a very casual approach to our excursions, selecting almost at random an adventure for the day. 'All this stuff is new to me too!' Nnamdi said. 'We are discovering New York together. I never paid attention to any of it.'

At the Metropolitan, after wandering through its extensive collection, we stumbled upon a touring exhibition of ancient Japanese wood furniture facing a garden of rocks, stones and fern trees built beneath a skylight. The garden exuded an almost palpable, serene beauty. The furniture was arranged in such a way as to align the viewer and draw them into the tranquility that emerges from a perspective that is intentionally limited, which is to say, focused.

A large placard said the guild of carpenters were so skilled it was an admission of failure to use a nail. We spent what seemed like hours seated on the aged furniture in the periphery of this garden, barely speaking.

As we were leaving the museum, my brother and I stumbled on an exhibition of Indian erotic carvings. On doors, in reliefs, enormous phalluses and endowed maidens copulated with bracing abandon. We spent an hour wandering back and forth through the collection. 'This is the life force of humans,' my brother declared, 'the sex impulse. And don't nobody tell you any different.' I did not agree, nor did I disagree.

Unknown to me, our time at the collection would trigger an idea in my brother's head. He would wander off to make or take calls as we walked to a restaurant where we enjoyed a late lunch. When we sat down at the diner, Nnamdi said: 'I presume you are still a virgin?'

'Yes, I am,' I responded and smiled.

'Well, I can't think of a better day than today to change that,' he said.

After lunch, he hailed a yellow taxi and told the driver to take us to the Sofitel Hotel on Forty-Fourth Street. At the desk he picked up a key and we rode an elevator up to a room.

'Have some wine, make yourself comfortable,' Nnamdi said. 'I have something special for you. I'll be right back.' I sat on the edge of the large bed and sipped red wine out of a tiny screw-top bottle I took from the small fridge. Then I picked up a miniature bottle of whiskey, unscrewed the top and threw it down my throat.

When my brother returned, he walked in with a beautiful young Japanese woman. 'Ichika is a college student. She loves New York.'

I stared. Ichika smiled at me. My brother placed a pack of condoms on the nightstand. 'Have fun, kids,' he said as he walked out of the room. I stared at Ichika. I could not believe how beautiful she was. Then she walked to the bed, stood above me and touched my cheek. Her hand was small, cool and soft. My nose filled with the fragrance of jasmine and lemongrass.

'Relax,' Ichika said, and then smiled.

The following morning my brother called Room 729 to summon me to breakfast. We ate together in the Sofitel's dining room and, at his invitation, I regaled him with an account of the activities. 'I got in. Then I moved it, back and forth,' I said, smiling happily at the memory. 'It really felt nice. We did it one more time before she left. The second time lasted longer. She said I was the best first-timer,' I boasted.

And my brother smiled at me. 'I am sure you are. Good work! That's how you keep your metaphysics warm!' He raised his palm up in invitation. I slammed it hard with a high-five.

'Now, I can give you a few tips. Show you how to satisfy a woman,' I said. And my brother laughed loudly.

12

PERPETUAL PERISHING PRESENTS

The Daughter of a Preaching Man

There is no such thing as absolute time; an event occurs only in relation to some other event. There is no such thing as absolute movement; a thing moves only in relation to something else. There is no such thing as absolute existence; a thing exists only in relation to the existence of something else. When one thing stands, another stands beside it.

Philip Bousquet was holding his wine glass up to the lamp, gently twirling the red liquid. The woman who would become my wife waited for him to continue, her left hand lying gently over mine on the table. Heidi's features, her long sharp nose and the raised cheekbones, glowed in the soft light. Her yellow hair cascaded to her shoulders, the strands carefully coifed in a tightly packed lattice that shimmered. I gazed at her irises often, mesmerised by the beautiful wavelengths which escaped. Her lips were painted a deep red. The strong smell of her rich perfume bathed and comforted me. I stared at her hand and its deep brown freckles against the white tablecloth.

We sat with Professor Rayburn and my future father-in-law at the back of the restaurant, and noise from the small tables in the foreground rose and fell in waves that intensified and dissipated irregularly. 'Your fiancé's theory is sublime,' Philip said. 'He established that time is not a fundamental property of the universe. The most profound implication of his theorem is that the lattice of space and time is entanglement. Now, it seems so obvious. But the truth is that before Kobidi published, no one in physics believed that any such thing was true.' Philip paused to take a sip from the glass.

'Before *The Length of Time*, nobody would even dare to make such a statement,' said Professor Rayburn, lifting a piece of fish on a fork to his mouth.

A waiter approached the table and tapped on a small white pad with a silver-coloured pen. Philip looked up at him.

'The wine is from the Sauternes region in Bordeaux,' the waiter said.

'Just as I thought,' Philip said, and grinned.

'I am afraid I am still a bit lost,' Heidi said. 'Does Ezeani's theorem tell what will happen in the future?'

'Well, in a sense it might be thought to do that,' Professor Rayburn responded. 'However, I think a more useful way to think about it is that it tells us, mathematically, if any length of time is *genuine*. If it belongs to the actual universe, present, future or past. The theorem is like a key that translates a type of encryption which tell us how the universe is entangled by analysing connections in spacetime.'

'You mean it tells us if the DVD is real or if it's a fake?' Heidi said.

'Exactly!' said Professor Rayburn.

'Well, I suppose it's a good thing I didn't lock you up, Ezeani!' the judge said, and slapped my back with his large hand. He laughed.

My future wife laughed and said, 'Dad!' Then she squeezed my hand and smiled at me.

'Ezeani knows I'm teasing, honey,' the judge said, still laughing. Heidi squeezed my hand again. My terror subsided.

'The fact that he was ever in danger of being locked up is crazy!' Professor Rayburn said, speaking in an earnest and formal voice. 'Kobidi is one of the most important and impactful theoretical physicists in decades.' His eyes were open and wide. He shook his head and then looked down at his plate. He looked like someone who might cry.

At that moment Heidi seized my face, turned it towards her and kissed my mouth. 'You are a star!' she said. '*My* star!'

'The thing to remember is that a length of time contains both space and time,' Professor Rayburn blurted to no one in particular. 'Time is a measure, a count, really, of the persistence of interaction. Of the endurance of entanglement, you could say.'

'God's work is amazing to behold,' the judge said. 'And his love is so abundant. Look at how much he reveals to you folks.' He stopped and looked across the table.

'In what way do you mean?' Professor Rayburn asked.

'You see, Professor Rayburn, I am a man of faith,' the judge said.

'Dad?!' Heidi said sharply.

'It's OK, honey. Professor Rayburn is just curious. Professor, you are right to question why Kobidi was arrested and brought before my court. It is indeed strange. But what you might not see is the Lord's will. Our Lord Jesus was himself brought before a magistrate.'

Philip Bousquet had put down his glass and looked across at Heidi's father. Professor Rayburn's eyes were fixed on the table.

'Heidi is my youngest child, and I don't think it's much of a secret she is my favourite. Unlike her siblings she didn't run off. She is a nurturer. And, truth be told, I worried about her spending so much of her life tending to me and the missus. It's God's wisdom, beyond human understanding, that Kobidi would be brought before my court. But if that hadn't happened, these lovebirds would probably never have met.' The judge turned to look at Heidi and me and nodded his head solemnly. 'Heidi has finally found what God meant for her. And I think their love has been instrumental in enabling Professor Kobidi to express himself. That love is the root of these profound insights you speak of.'

Professor Rayburn looked up from his plate. It appeared that the judge's words had cheered him up. He had a smile on his face. 'You think God was playing matchmaker?' he asked and then beamed a wider smile.

This dinner took place on the evening of my appointment as a tenured professor of physics at Cornell University. At this time Heidi and I had been engaged for almost a year. She selected the restaurant and arranged the invitations to the small group with the same efficient and assured energy she had taken to organising and directing my life. Heidi protected me, fought for me, and enabled me to focus on my work in physics; she sheltered me from the confusions, insensibilities, madness and moods that accosted me. And when the absurdities and confusions threatened to overwhelm me, Heidi would hold and comfort me until the anxiety and terrors passed, like dissipating waves.

On a crisp day, several months before I would be granted tenure, we ate lunch on a blanket in the Quad and Heidi raised a napkin to wipe ketchup from my cheek, then told me it was absurd for a professor at Cornell to continue to live at Telluride. This thought had never occurred to me. I had never considered where I should live, or the appropriateness or absurdity of one place in comparison to another.

'I think we should get a house. There are a few places I have shortlisted. We can go and see them tomorrow afternoon,' she said, and smiled gently at me.

We looked at three houses. When we walked into the third – the five-bedroom house behind the trees – I was overwhelmed with the feeling of home. It was like I had come upon a place in which I had once resided. This feeling was immediate and overwhelming. Heidi sensed my emotions and beamed. 'I thought you would love it too! Yes, this is home.' She squeezed my hand and then kissed my lips. The realtor, still in my line of sight, smiled and pulled a laminated folder from her large handbag. We wandered through the many rooms: the den, kitchen and study on its ground floor, the master bedroom and three bedrooms on its second floor, the small bedroom with sloping walls in the attic – converted, I was told, from a crawl service space. We came down the stairs onto the backyard porch on which I would, on a bright afternoon, many years in the future, receive an epiphany, and understand the true meaning of entanglement and the import of my life's work.

I would live here till the end of my natural life. In those years, I was occasionally struck by the ways in which this house behind the trees seemed to be an amalgam of the two houses on the campus at Ibadan. There was the semicircular driveway that came through the hedges and trees; and beyond the back porch, the raised stone wall that led to a slope of rocks and boulders; and the green trees and grass that seemed to enclose the house and set it away from the world. As in the first house on campus in Ibadan, it was on the rocks that I would first observe Anyanwu. Only much later would I discover that he had moved into the room in the attic, coming and going as he pleased.

Before the end of that week, Heidi and I walked into a lawyer's office and executed papers that established purchase of the house on Hanshaw Road. As I signed, I noted the large sum paid for the house. A number so large that I could not really comprehend it in any way consistent with my own experience of money. It is a measure of the ways in which Heidi secured and protected me that I did not have any use for this information.

Our engagement to be married was announced that afternoon in large paid advertisements in the *Ithaca Journal* and the *Cornell Daily Sun*. In its weekend issue, the *Ithaca Journal* ran a piece about us. In it we were referred to, several times, as 'the couple'.

The restaurant at which we celebrated my appointment was located in a large house tucked down a lonely, wooded street on the edges of Ithaca. When the waiter produced a cheque in a black leather folder, Heidi tried to pick it up but her father grabbed it and would not listen to her entreaties that he relinquish the bill. The judge produced a shiny silver card and placed it into the folder.

At the door, Professor Rayburn grabbed his coat from the hands of an attendant. He rushed out without bothering to put it on. As I slipped my arms into the dark-grey cashmere overcoat Heidi had purchased for me, her father held out her own and she placed her arms within it.

Snow was falling lightly, and the air was crisp, cold and fresh. I could hear bird calls. The judge stood beside our car, kissed Heidi on each cheek and then shook my hand. 'I am proud of you, son,' he said. Heidi drove our large car out of the small parking lot onto the black road that glistened in the headlights' beams, waves of light, refracted by the thin layer of water molecules floating on the melting snow.

As she pushed open the door of our house, Heidi threw her keys in a bowl on the shelf in the front hall. 'Well, Professor

Kobidi,' she said, 'welcome home!' She pulled me to her and kissed my lips as she shut the door with the heel of her foot. Her rich fragrance enveloped me.

An Interview with My Sister

The black town car pulled into the short driveway. My wife reached for my hand and squeezed. As I placed my foot on the ground and pushed my head into the cold air, I saw my sister standing at the door in medical scrubs, a cream cardigan around her shoulders.

'Welcome, Ezeani,' she said, and then turned back into the house. I could hear the crunch of Heidi's feet on the gravel. She held my hand as we climbed up to the porch together.

I knew Obiageli was a physician and my assumption had been that she was wealthy or at least well-off. In the years I had lived with Heidi, I had developed a general sense of the material value of things like furniture. The objects in my sister's house did not have the solid, substantial feel I had come to associate with the homes of my wife and father-in-law's set. My sister's living room was filled with mismatched things and had a temporary and improvisational air. Like the work of an amateur.

There were two large sofas, each too big for the room, suffocating the space. Against the far wall was a tall shelf and over it hung a large, framed picture of my mother. The living room floor was covered by a shaggy brown carpet with a colourful heap of plush puppets and a few plastic children's toys in a corner.

Heidi and I sat on a large couch, drinking tea. Obiageli on a straight-backed chair facing us. Heidi and my sister were discussing children. I stared at the photograph of my mother.

It stared back at me, with a smile. The smile frightened me, and I looked away. My sister had a child. She had been married briefly. Her ex-husband was mentioned, and Heidi made a kind remark as if consoling one robbed by an unapprehended criminal. Heidi was expressing our wish to have a child of our own. 'We are really looking forward to becoming parents. Unfortunately, nature is not being entirely cooperative,' she said. My sister nodded sympathetically. I looked at her eyes. I could see she could continue this conversation with Heidi for a decade without ever putting any of herself into it, like she was auditioning for a part in a boring recital. Heidi, occasionally, would turn her eyes to me, as if willing me to speak.

'I am really glad you are in Ezeani's life,' my sister said, suddenly earnest. Her words were directed at Heidi but her eyes were fixed on me. 'I am not sure how much Ezeani has told you, but we grew up in a difficult home; it is a literal miracle that we survived. It is a miracle that he is here.'

Heidi looked at me, perplexion on her face. 'His description of your childhood in the village and Ibadan seemed almost idyllic, with Anyanwu providing magical colour.'

A small, wry smile passed my sister's lips. 'No, it was not idyllic.'

'My father's brother lived with us for some time,' I blurted. 'He was not always nice.'

'He was a paedophile,' my sister said. 'He raped me.' Her eyes squinted, then focused and fixed on me. I turned my eyes to the carpet. Heidi was rising and reaching an arm out to my sister. I could feel her shrug Heidi off. Then I heard my sister's voice, low, clear and distinct: 'I told our father. He said nothing. He did nothing.'

'Honey, that is awful. Perhaps your father did act? Kobidi has told me so much about him. I can't imagine he would not have protected you,' Heidi said, leaning towards my sister.

'He did nothing,' my sister said, in a shallow, flat voice.

'Dear, how can you be certain.' Heidi spoke in a flat tone that did not have the raised inflection of a question.

'Because one day he tried to rape me himself,' my sister responded, her face fixed on Heidi's as she rose. I could feel Heidi's hold on my hand loosen. Then, surprisingly, my wife raised a hand to her mouth. I saw wetness start to gather in Heidi's eyes. She lifted her other hand also to her mouth. This shift in Heidi's sentiments surprised me. I wondered what caused my sister to hate our father; what induced her to tell such hurtful lies.

My sister turned her back to me. She walked over to the large shelf and pulled an enormous bundle of letters, tied together with brown twine, from a drawer. Holding the twine between her fingers, she dropped the letters beside the teapot. I immediately recognised the handwriting. It was my impatient hand.

'I am not sure you can stomach the awful, hateful things that are written here,' Obiageli said, pointing at the letters. 'They aren't really Ezeani's words. It was our father, subtly filling his mind with bile.' I could sense the pressure of gas fermenting in my belly and taste its bitterness in my mouth.

There are only two fundamental forces – love and strife; the force which binds things together, and that which tears things apart. The two are hopelessly linked.

Brother, Can You Spare a Dime?

Nnamdi's phone calls had become more persistent and difficult to follow. It was Heidi who suggested I stop answering them.

The substance of his trouble had been the subject of an article in the *Wall Street Journal*. The acquisition of the investment bank my brother worked at was followed by claims of sexual harassment against some of the highest income generators at the firm, the most prominent of whom was my brother.

'It's all just politics,' Nnamdi explained early in the scandal. 'The new corporate overlords are looking for leverage; trying to prevent us from leaving.' He was laughing when he got off the phone. There was a tone in his voice that made me think that he had been crying.

These conversations coincided with my appointment to the faculty of Cornell University and my receipt of tenure. It was a period of purposeful activity; the period in which Heidi helped me shape my life. I was increasingly unable to put Nnamdi's troubles in mind and follow the contours of his travails.

One afternoon, my brother, unshaven and dishevelled, rode up the drive of the house on Hanshaw Road as a passenger in Chijioke's grey minivan. The visit was unannounced. I was in my study, staring out the window. Behind me was the large blackboard I had installed across one wall. A single diagram of three interlocking squares was drawn on it. I heard the front door open and Heidi step out and say: 'Well, hello! Was Ezeani expecting you?'

It was summer and the air was only just starting to cool. Heidi served a late lunch on the small wooden table out on the patio at the back of our house. I lifted a fork laden with roasted chicken to my mouth.

'The case against him is civil. No one is even talking about anything criminal. There is no evidence. Of course he has countersued, but where does he go to get his reputation back? He is an African man accused of sexual misdeed by a Caucasian woman. It's a ridiculous railroading,' Chijioke said. She placed her hand on Nnamdi's arm as she spoke. Heidi said nothing in response, but she made an ambiguous sound. Chijioke kept talking.

I could not follow all of it, but I gathered that Nnamdi had at some point run out of money and was suing the law firms that represented him, along with his former employer. Chijioke was his current lawyer. Nnamdi spoke very little.

It was after dinner, when darkness had descended on Ithaca and we were making our way to the minivan with Heidi and Chijioke walking a few feet ahead of us, that my brother held my arm to detain me. 'I am really going to need a loan, Ezeani,' he said. 'If you can spare fifty thousand dollars that would really help.'

I stopped and stared at him. It was such a large sum of money. 'I don't have fifty thousand dollars,' I responded.

'Of course,' he said, 'but you are married to a multi-millionaire. Her family is one of the wealthiest in New York. Just mention it to her. I am staying with Chijioke in Syracuse for a while. Lost my place in the city.'

Later that evening, hours after my brother had driven away in the grey minivan, Heidi pushed open the study door. I was staring at the blackboard. I turned to her, slightly disoriented.

'Sorry to disturb you, but I think it is important I say this,' Heidi said. 'You need to be careful with this situation with Nnamdi. It's very public and the media is really invested in the story. You don't want the taint to extend in any way to you.'

One cannot understand life without understanding time. As Heidi spoke, my eyes were filled with a vision of my brother's face as he lay in the bed beside the door in the room we shared in the first house on the campus of the university at Ibadan.

I continued to stare at Heidi. 'I really don't understand. Nnamdi said he needs fifty thousand dollars. Handle it the way you think best,' I turned to look at the blackboard.

'Good night, my love,' Heidi said, stepping further into the room and kissing the crown of my head.

As she closed the door, I closed my eyes, and tried to see time.

PART III

13

THE VARIETIES OF MARITAL EXPERIENCE

The Sacrament of Marriage

Outside the tall windows in the living room, the sun was slowly setting, imparting an orange glow of refracted light as it moved across the sky, away from me, towards the horizon at the curve of the world. I was sitting still, in an armchair, in the living room of the home I shared with my wife. But I knew this wasn't really true; I was also travelling about a thousand miles per hour, carried on the Earth, spinning away from the sun. All that could truthfully be said was the sun and I were moving, one relative to the other. Nothing moves except in relation to something else.

Heidi's frame moved across the window. Momentarily, she blocked the light as she laid a tray with a teacup and a slice of carrot cake on the console table beside me. There was a smile on her face. As she walked away, I spoke.

'It's actually common practice among the Igbo. Traditionally, the male head of household has his Obi. It's a space in which he receives visitors, says prayers; a private den, secluded from his wives and domesticity,' I said. I picked up the fork and sliced into the carrot cake, then lifted it to my mouth.

'That's what the Upright Lodge that Chi built is called? An Obi?' Heidi asked.

'Yes. Chi is the male spirit of the Parent Gods – Chi and Eke. The Obi, you could say, represents the desire for male independence,' I replied.

'Male independence?' Heidi asked, her brow slightly furrowed, but a smile playing on the corner of her lips.

'Yes,' I said, and I beamed at her. 'In fact, in the Igbo formation narrative, this desire for independence is what leads to the manifestation of the universe. Chi starts it all when he builds his Upright Lodge and excludes Eke. It's what causes the Big Bang!'

Heidi laughed. 'That's such an interesting concept,' she said. 'The universe exists because the male spirit needs a man cave.'

'And the female spirit won't abide it,' I said, laughing a little. Heidi chuckled. 'We laugh,' I said, starting to smile, 'but it might actually offer a clue to one of the biggest unanswered questions in physics: the vexatious problem of matter/antimatter asymmetry in our universe.'

'I have no idea what that means,' Heidi said as she sat down in the armchair across from me.

'It's essentially the question of why there is something instead of nothing. The universe should have matter and antimatter in equal quantities and they should annihilate each other. But that's is not the case. Antimatter is missing.'

'So it's hiding in Chi's Lodge?' she smiled, before she lifted the glass of red wine to her lips.

'No one seems to know where the antimatter is. Perhaps I will look there,' I said.

Heidi laughed, then she drank again. She put the glass down, fixed her eyes on me. I stared at her beautiful eyes, and then her body as she rose, walked over to me. When she bent over me, she kissed me fully on the lips.

'You are such a brilliant man,' she said. Then she kissed me again. She straightened herself and patted the hips of her green dress. Leaning over, she dimmed the lamp on the side table. Heidi knelt between my legs, kissed my lips, then my neck. She nibbled at my ears. I could hear my breath thicken. I grabbed the ends of upholstered armrests. She fumbled with my belt. I closed my eyes.

'Symmetry is beautiful, it is pure in ways that is hard to even describe. It's what drives equations. Asymmetry is ugly, crude and inelegant, but it is why the universe and life exist,' I whispered.

Heidi had loosened my trousers. She did not appear to be listening to what I was saying.

Native Son

Can you shine two beams of light at the same spot and get darkness? If you understand physics, then you know that the answer is: Yes. Two waves of light can interfere with each other. If the crest of one wave coincides with the trough of the other, in a process called destructive interference, darkness ensues.

Our efforts at natural procreation yielded nothing.

After several years of effort, our son, Njoku Albert Kobidi, conceived through a process of in vitro fertilisation, was born at the NewYork-Presbyterian/Weill Cornell Medical Center in New York City on a bright day in the middle of August. I attended the birth.

In a coffee shop across the street from the hospital, I sat at a wooden table beside a window that looked out into the street, and drank orange juice, occasionally nibbling on a scone. The cup of coffee I had purchased cooled. I watched the steam rising

225

from the white mug until there was none left. The presence of heat is the only distinction between the future and the past. Hot things cool. That is how we know where the future lies.

The notebook I now carried everywhere lay open on the table. As I flipped to a page in which I had written down vector calculations, I noted a scribble in the margins. It was in Anyanwu's large, uneven hand: *What are children for?* I scratched out the note with my pen, drawing, and then painting in, a blue rectangle.

A few days later, travelling in a chauffeured van, my wife, our infant son and I returned to our home in Ithaca. During our absence, at the instruction of my wife, the baby's crib had been installed in our room. Another room, its walls covered in a lovely blue wallpaper, had been converted into a nursery for our son.

The nanny, an older woman in a black and white uniform, was waiting as we came up on the porch; my son, wrapped in blankets in my wife Heidi's arms. The nanny threw out her arms to receive the baby. 'No. Not right now, Juanita,' my wife said, as she turned her head and smiled at our child.

Later that night I was awoken by a small, shrieking cry. It came from the crib on my wife's side of the bed. At first, I was confused. The piercing sound gathered intensity, turning into a wail. Heidi was fast asleep, gently snoring. I raised myself and walked round the foot of the bed. The neonate was now bawling loudly. I was unsure of what to do. I looked at my wife. She seemed so at peace. Then I picked up the child, as gently as I could. He stopped crying for a moment, as if assessing his changed circumstance, and then he seemed to gather enormous energy and started wailing even louder. Heidi sprang up in the bed, startled and confused. 'What's going on, Ezeani? What are you doing?'

'The baby is crying,' I said. Heidi jumped out of bed. She

held out her arm and I passed the baby. 'He must be hungry,' she said. 'I slept through his feeding.'

My wife arranged her body on the rocking chair in the corner of the room and started feeding the baby. In the glow of the nightlight, I could see her cooing and whispering to the child we had conceived with my sperm and the egg of a donor that had been selected on the criteria of hair colour, eye colour, and whatever marks of intelligence we could glean from the Bio WorkUp profiles that my wife had spent hours reviewing, then winnowed to a shortlist of three profiles that she placed before me. 'It is down to these, Ezeani. I need you to make the final selection. It needs to be your decision too.'

I did not know on what criteria to judge. Certainly, I gathered from her comments that my wife desired the donor to look like her. But all the candidates vaguely and superficially resembled her. I stared at the profiles for a few minutes, unsure of how to proceed. Then a sudden, logical clarity came to me. I chose the youngest.

'The problem with her is you don't see drive in her profile. She does not seem like a go-getter,' my wife said.

Conversations in the Dark

The light from the landing fell through the open attic door. My wife was standing in relief, her arm stretched across the opening. There was a haze in my head. I was lying on the bed, my head propped up by the pillows, and like a person emerging from a dream I tried to comprehend what was going on. Gradually, I realised I was shouting. I listened to what I was saying. It was a loud repeated screech: 'LEAVE ME ALONE!'

'What do you think I want from you, Ezeani? What do you imagine I am after?' Heidi said.

I did not speak.

'We built a life. We have been able to do so much. You are a brilliant man. You are doing wonderful work. God knows, you may even be able to get a Nobel. Phil is convinced of your talent. As long as you let me help you, it is possible.'

'Phil? Why are you talking to *Phil*? What are you discussing with him?'

'What do you think I am *discussing* with him? Don't be absurd! You need help. You won't accept it. I don't know what else I can do.'

'Leave me alone,' I whispered, turning away from the door.

'You lock yourself up in the attic for days, you won't come out, won't take a bath, won't answer the door. You are banging on the floor, shouting, knocking on the walls. Threatening to burn everything down. It's terrifying! Njoku is frightened by the noise.'

I nodded my head. I was waiting for her to finish talking, but the stream of words continued. Finally, I shouted: 'I KNOW YOU ARE TALKING TO THEM BEHIND MY BACK. LEAVE ME ALONE!'

I closed my eyes. Her voice did not stop, it was in my ears, the words pouring out. I only opened them again when the words in high register stopped and I heard Anyanwu say: 'She's finally left. Sit up. Let's drink something.'

The (Real) Thought Police

In the four years and three months before my death, I was twice confined, on an involuntary basis, to the Richard H. Hutchings Psychiatric Hospital in Syracuse, New York – on each occasion for two months. The first confinement was a year, six months and six days after the birth of our son.

On both occasions, I attempted, unsuccessfully, to secure my release by presenting legal arguments before a magistrate.

My confinement – on each occasion – was triggered by a petition filed by my wife Heidi to judicial officers of the State of New York, supported, as required by the Mental Hygiene Law of the state, with the certification of two physicians to the following purported facts: (1) my judgement, as a result of my mental illness, was too impaired for me to understand the need for care and treatment in the psychiatric hospital; and (2) I posed a substantial threat of harm to myself or others. Although they made reference to the possibility I might harm myself, the petitions by my wife seemed focused on the ridiculous suggestion that I might harm her or our infant son.

I was lying in the attic room on the top floor of the house I lived in with my wife on Hanshaw Road – my feet, with my winter boots still on, resting on the bed. The walls of the room had been decorated with a grey-green patterned wallpaper. It was an early winter dusk, the darkness settling in the later afternoon.

Anyanwu was seated at the small desk in the corner of the room. He was dressed in the white, red and black cloth, the ends gathered and thrown over his shoulder. A stiff red cap with a large white eagle feather was fixed on his head. He was rifling through my papers, skipping through the journals, briskly flipping the pages as if he was looking for something. Then he turned sharply to me.

'You didn't say anything about me!'

'They are scientific papers.'

'I have read other scientific books. The authors mention their God sometimes. You didn't mention me even once?!' Anyanwu was speaking loudly.

I was tired. He looked at me, then his eyes softened. 'There is still time,' he said. 'Don't worry, Ezeani, there is still time.'

I rose from the bed, walked to the desk and pushed aside the papers that he had scattered. Two notes on the open page of my notebook caught my eye. 'What is it?' Anyanwu asked. I lifted the notebook and walked back to the bed, lay down and stared at the page. The notes, written in my hand, read:

Cause and effect are illusions. Unless there is a burnt sacrifice the world flows both ways.

And:

The past is determined by the present neither more nor less than the future is determined by the present.

I closed my eyes. After a few moments, I heard a few sharp knocks on the door. 'Open the door, Ezeani. Please. Please.' It was my wife Heidi's voice.

'Leave him alone!' Anyanwu screamed. Then he turned to me and laughed. I could sense myself drifting away from consciousness. My time stopped.

When I became conscious again, I was unsure how much time had elapsed. Was it days, weeks, hours? I could hear a muted din in the background. It held my attention. I could not recall where I had heard the sound before. As the moments passed it became the clear sound of multiple sirens.

'Get up! Get up!' Anyanwu said, shaking me roughly by the shoulder. He pulled me off the bed, almost dragging me to the window. 'Look, they are coming into your driveway.'

Outside, in the dimming light – was it the same day as before? – I could see an ambulance sandwiched between two police cars. Despite the sirens and the flashing lights, the convoy of vehicles seemed to be moving at a deliberate, unrushed speed up our long driveway. Anyanwu stood at the window, his hands resting on his hips. He was wearing nothing but the loincloth around his waist. His chest muscles rippled. He turned to me. 'They are here to get you,' he said. 'Maybe there is a place where you can hide. Perhaps, under the bed.'

My head fell. I went to the bed and sat down. 'No, I meant hide under the bed,' Anyanwu said. I ignored him. Anyanwu picked up the chair and small desk and placed them against the door to create a barrier. There was a knock. Then a male voice said, 'Professor Kobidi, please open the door. We are with the Ithaca Police. We need to have a conversation with you.'

'Get out of here before Amadioha strikes you down!' Anyanwu yelled at the top of his voice. He was screaming in Igbo, yelling curses at the man on the other side of the door. When they started battering the door – kicking it, I presume, with their feet – Anyanwu switched to English. 'You motherfuckers better get the hell out of here! I will sue you for every cent you have, assholes.' The battering sounds continued. Anyanwu came over and sat beside me on the bed. He put his large hand on my right knee. Then he said: 'They will be able to get through. Don't tell them what we discussed. They cannot hold you forever. You are being tested. Be stout-hearted.' Then he blew a froth of saliva in my face and my time stopped.

When my time started again, I was looking up at the roof of the front porch of the house on Hanshaw Road. The air was cold. For reasons that were not entirely clear to me, I was naked. A man was pushing me down on a gurney. His hand was strong and firm. His eyes seemed soft, and glistened. As I attempted to rise, he pushed harder on my chest. The eyes were now fierce. A woman appeared beside him, and they worked together to fasten me to the metal bed with a thick black polypropylene belt which had a metal clasp attached to a lever. The lever made a clicking sound as it was tightened. It must have gears, I thought.

They were both wearing red and black coats with a square badge on the left chest. The badge had a red cross with a viper wrapped around it. When the woman leaned down to throw a blanket over me, the viper opened its mouth and spoke to me: 'Anyanwu ran away when the police came! You deserve a stronger God. Follow me instead!' There was mockery in the snake's voice. It licked its thin lips with a forked tongue, then it hissed. I was terrified. My body shook. I tried to draw breath in the cold air.

They lifted the gurney and pushed it, on rails, smoothly into the ambulance. Heidi's face appeared in the frame of the ambulance's door. She was speaking to the man. 'Is he OK? Should I come in the ambulance?' my wife asked. The man shook his head. When they both turned to me, I could see the canine teeth flanking their incisors, which were elongated like the viper's. I screamed. My time stopped.

It is a strange fact that my wife's actions never appeared to me as a betrayal. I believed them explained by her lack of comprehension of the new physics of entanglement I had started working on. And, I suppose, her utter confusion as to the basis of my conviction concerning the appropriateness of my conduct. Conduct that I freely admit must have appeared confusing and upsetting to her. In her typical efficient and practical frame of mind, she must have resolved that the steps she took were necessary to ensure my return to what would appear, to her, as productive physics and conventional conduct. I rejected my brother's suggestion of a more sinister purpose.

Strangely, what felt like betrayal was the fact that Dr Chan

— whom I had first consulted as a condition of my release from detention on a material witness warrant, and was a man with whom I had shared some sincere observations about life and science and whom I believed had a sympathetic, if simplistic, appreciation of my approach to physics — was one of the physicians that executed my confinement order.

On both occasions of confinement, my petitions for release were filed, after rushed consultations by Chijioke, at my brother's instigation. The petitions failed. I was only able to secure my freedom from this prison — there is no other word for it — when, on Anyanwu's advice, I renounced my 'delusional hypothesis', abjured Anyanwu himself, and repeated the conventional description of reality that I was aware the psychiatrists required of me. On both occasions, this strategy, as Anyanwu had predicted, almost immediately secured my release.

In the period between these two spans of confinement, a period of surveilled rationality, I was able to produce, in addition to my final theorem, two strong papers on matter/antimatter asymmetry, one working with Phil Bousquet and Lawrence Alcott, and the other a paper I co-authored with Milford Herbert at Caltech, titled: 'Charge Conjugation: Existence and Time Asymmetry', which in addition to establishing the validity of the Kobidi-Herbert Function, set out a strong theoretical framework for analysing, as Herbert put it, 'the unevenness in the universe which makes time move forward'. It was good work, of a very conventional aspect and with sound mathematical foundations, but it contained nothing of the new insights on entanglement that had started to consume me.

After my second period of forced hospitalisation, I adopted Anyanwu's advice wholeheartedly and kept from others, especially my wife Heidi, my real thoughts. I was also careful to adopt relatively moderate behaviours. In this way I avoided hospitalisation and the attention of psychiatrists until the end of my natural life.

The profound and beautiful understanding of our entangled existence that I share with you, I had shared in its nascent forms with Dr Chan. It was these insights that Dr Chan, a man with little more than a rudimentary understanding of physics, had labelled a 'delusional hypothesis' to the judicial officers of the State of New York.

It is Natural That Love is Blind

Soon after we were married, my wife discovered that her options for vacations were to be severely limited by her selection of a spouse. I had returned to the house on Hanshaw Road to find her in the living room looking through thick travel brochures. 'Hi, Ezeani! Where would you like to go on vacation this year? Italy? Spain? Morocco?' she asked as I entered the room. 'Please don't say Italy. I am tired of Italy. We went almost every summer when we were kids. My father was obsessed.'

'How will we get there?' I asked.

'On a plane, of course,' my wife said, smiling.

'I don't want to get on a plane,' I said.

'How can you get anywhere if you don't get on a plane?' Heidi said, laughing. 'Wait, are you serious?'

'I have only been on a plane once. Getting to America. I was terrified. I will not do it again.'

'Was it a turbulent flight? What happened?' Heidi's voice was softer, and there was concern on her face.

I told her about seeing the moon over the Atlantic, large and grey, alone in the sky; my terror when the storms started and the plane shook, the flatulence that filled the plane's cabin as I held on to my seat's armrests with both hands and then closed my eyes.

Heidi walked over to me and put her arm around my shoulders. 'Sit down,' she said, as she led me to an armchair. After I sat, she lingered, perched on the arm of the chair. Then she leaned in and kissed me. 'Have you ever discussed this with Dr Chan?'

I shook my head.

Eventually, Heidi accommodated herself to my refusal to step on an aeroplane. We started our tradition of taking our holidays in the beautiful retreats available by road, mostly on the Eastern Seaboard of the United States and Canada.

Our most frequent vacation location was a beach house on Martha's Vineyard that we rented for six weeks at a stretch for six years in a row. Heidi would plan the vacations in great detail, arranging for a rotating roster of our acquaintances, my colleagues, her friends and family to spend time with us in the large house. We would drive east towards Boston and then head south to a small town called Woods Hole, from where we would take the ferry to the island.

The house we rented was large, larger even than our house in Ithaca. It had eight bedrooms, a considerable number of toilets and bathrooms, and a deck that wrapped around the entire house. The exterior was clad in grey weather-beaten shingles that concealed the beauty and comfort of the spaces within. We would spend the first week of our vacation alone in that enormous house. Heidi would coax me on a daily evening walk around the island. On the way back, our arms entangled, we would often stop and buy ice-cream cones from the vendor at the top of the hill, on the corner that led, down a steep incline, to the rented house. There were always children gathered around the pastel-coloured stand and my wife and I would chat while we waited our turn.

In the first week, Heidi prepared our meals and we would eat, side by side, at the large wooden table as we talked and laughed. She would sip red wine while I drank orange juice. My wife would tell me stories about common events that were in the news. Many of them amazed me. Heidi often laughed at how cut-off I was from the quotidian beat of life. Sometimes we spoke about my colleagues in physics and which of them was doing innovative work. My wife managed my correspondence and helped me in selecting, contacting and arranging to work with the most promising people. She helped enormously in focusing my efforts and I reaped the reward in the many collaborative papers I worked on in that period. None of this work had the brilliance of the initial papers I had completed before my PhD, but it was definitely good, respected science. It was a period in which I settled into a warm, familiar comfort and did my work in a measured and perhaps plodding way. I was still convinced of the genius of my insights and I got encouragement and pleasure from the frequency with which my first three papers were cited. I had lit the first fires, and there was promise that I would continue to deliver seminal breakthroughs to our science. There was an air of anticipation around me, and, balanced with these various collaborations, I was convinced that I would soon move in a new, even daring, direction. During this period, I lost touch, for several years, with Anyanwu.

In the second week, our domestic help, two women named Mariana and Rosa, would appear. They were responsible for preparing meals and cleaning the house for the rest of our summer stay. In the fourth year at Martha's Vineyard, the women were Rosa and Guadalupe, and for the two years thereafter it was Rosa and Alejandra. The women assisted us and our various guests by cooking, serving, fetching and cleaning.

Everyone spoke to them politely as they asked for these services, always adding the servant's first name to each request.

A day or two after the domestic servants, the guests would start to arrive. Usually, in the second week, my wife's family would come. Her father and mother arrived first and stayed longest. Then her two siblings and their children. Scheduling their visits was always a point of some conflict, a function of the complex work schedules of her sister, a cancer doctor at the NewYork-Presbyterian/Weill Cornell Medical Center, and her brother, a lawyer in the employ of the United States government in Washington, D.C.

During the day, my wife, her siblings and their children would leave the house and embark on bike rides and swimming excursions. When everyone left, I would sit on the porch looking out towards the water, a stack of books unread by my side.

One morning, my wife's father came out on the porch. We sat in silence for an hour or so and then he spoke.

'Professor,' he said, turning to me with a sigh, 'what does Einstein's theory of relativity really mean? I read somewhere that it shows there is a God by scientifically establishing a genesis – the big bang.'

I hesitated and then offered a brief explanation. 'There is no experiment that you can conduct which will tell you whether you are moving,' I said. Then after a moment I added: 'This suggests movement is not real. Einstein was the first to clearly see this. In the new work I am doing I follow that train of thought. If movement is not real, then perhaps the medium in which movement occurs – what we call spacetime; maybe space and time are not real.'

The judge laughed and took a sip of his lemonade. 'I don't mean to laugh, Professor, but I am having a hard time getting my mind around this.'

We lapsed again into silence.

'If you can get me to the essence of it, without all the math. Just the main idea,' the judge blurted.

'Judge, you need to let go of the world your common sense tells you is there, to see what Einstein meant clearly. Let me try to explain. If you can only be said to be moving in respect to something else, some other object, then there is no absolute space against which you are moving. Do you see this?'

'OK,' said the judge. 'I can concede that. But how does that mean there is no time?'

'Because space and time are dimensions of the same thing. Imagine the space inside your body: when your body moves relative to my body, that space is moving relative to the space in my body, and if the space in your body is moving relative to the space in my body, so is the time in my body moving relative to the time in your body. There is no absolute space and no absolute time. We just need space and time to connect our relationships to each other. Otherwise, space and time is super-fluous. That's the heart of it,' I said.

'I think I have a sense of your drift. Run that by me again, Professor,' the judge said.

'Stop bothering Ezeani,' the judge's wife said, returning at that moment with the rest of the swimming party in sun shirts and damp swimming suits, with colourful towels wrapped around their shoulders or in cylindrical bundles under their arms. 'Alejandra dear, please can you come grab all these towels from the kids? They need to be in the wash and dried for tomorrow.'

In the last week, when the guests were primarily my colleagues, the evening dinners would run into the early hours of the morning with sublime discussions of physics and sometimes, heatedly, the politics of the science. Heidi stayed late listening to these conversations, her hand over mine. No one rose early the next morning.

On what would be our last summer on Martha's Vineyard, Philip Bousquet arrived at the house with a short man by his side. The man was wearing thick black spectacles. He stood awkwardly in the door frame, his travel bag in his hand. There was something familiar in his face but I could not immediately place him, not until Heidi got to the door, kissed Philip on the cheek and declared 'You made it!' then hugged the little man. His eyes scanned the room, and when he saw me rising from a chair on the porch he ran to me and hugged me tightly. 'Kobidi! Floreat! How great to see you!' Then he turned to Heidi and said: 'You've made a very respectable academic of this rather unfocused mathematician,' and laughed. It was then that I recognised my first friend, Ope Adesola.

'Adesola!' I screamed. Then I held out my hand and he shook it.

Bousquet, Adesola and three other physicists – Kevin Scanlan at Caltech; Brad Wheeler at the Institute for Advanced Study; and Georg Hess who was visiting the United States to help his son settle into school, far away from their home in Bern – were gathered at the table. They would spend the last week of our last summer in the house at Martha's Vineyard with us. It was a stimulating, wonderful week.

'Yes, I have known Kobidi for almost all of my life,' Adesola said. 'We went to the same primary and secondary school, and then we were undergraduate roommates.'

'Philip tells me you are doing a sabbatical at Yale. What's your area?' Scanlan asked.

'Developmental anthropology,' Adesola said.

'Like the anthropological study of infants?' Scanlan continued.

'Not really. How developmental aid and other efforts can be better tailored to help countries in Africa and Latin America. I sort of drifted into it,' Adesola responded.

'Anthropological study of infants sounds more interesting,' I said, and smiled.

'I suppose. They might be more similar than you think,' Adesola said, barely smiling.

After dinner at the large table, we slowly drifted to the living room. As we settled in the deep couches and wingback armchairs, my ears picked up, in the background, the small clicking sounds of the maids clearing the table. When Wheeler started laying out the algebra underlining his work on gravitational wave propagation, Heidi squeezed my hand, excused herself and walked, in her bare feet, to our room. Hess stood up when my wife rose and pushed his head down, as if making a bow.

14

RELATIVITY AND THE PROBLEM OF SPACE

The Critique of Pure Reason

Reality is not what you think it is. Anyanwu taught me this when I was still a child. Before I learnt to clean my teeth.

My study of physics confirmed that Anyanwu's self-serving, disjointed and misleading assertion was, at its core, true. Physics' clear mathematics, experiments and reason assured me I was not deluding myself – reality was not what it appeared to be.

In the last of the three beautiful papers I wrote before I earned my PhD, I included a prologue that contained the following quote:

> It is characteristic of Newtonian physics that it has to ascribe independent and real existence to space and time as well as to matter . . .
>
> Albert Einstein, 'Relativity and the
> Problem of Space'

This is the thing that Anyanwu taught me; it lies at the heart of what I need you to see – there is nothing independent or real about space, time or matter. The only things we have are our relationships. The relationships create us; they also define us. The term we use in physics is *describe*; this, perhaps not coincidentally, is also the word novelists, screenwriters and playwrights use. Our relationships describe us.

At the time of my death it was this work from my early papers, supported, to a lesser degree, by my somewhat pedantic and more conservative work on matter/antimatter asymmetry, which had formed the foundation of my reputation in physics.

However, the most impactful work of my natural life — the work that linked our understanding of Einstein's relativity with the confounding absurdities of quantum field theory and laid out a comprehensive theory of our universe through a correct appreciation of the true nature of entanglement; the work that would cause me to cry at the sheer beauty of its underlying mathematical structure — this work was derided by many of my colleagues, and abuse was heaped upon me.

The irony of this turn in affairs is not lost on me. It was of course inevitable; I can see now that it could have been no other way. The battle between potential and becoming, between seeing and being seen, required it; as did the battle between Anyanwu, his unexpected cohort of allies, and the emergent Empress Earth Goddess Ani; the battle between my wife and my brother; and the battle between my living father and my dead mother. But I did not, I must admit, comprehend this as I lived the moments of my life. I was blinded. I did not see clearly the connection between the thing that stands and the thing that stands beside it.

What we know is eclipsed by what we feel; in fact, properly understood, what we know is actually only the most attenuated and precisely calibrated form of what we feel. I had to abandon my slavish adhesion to mathematics to see this; perhaps it is more accurate to say that I had to put aside my presuppositions on mathematics and logic to apprehend it, just as, as a young student of mathematics, I'd had to put aside the rational geometry suggested by what passes as reality to comprehend imaginary numbers.

And what I saw as I questioned time was that the mathematics that delivered imaginary numbers also implied imaginary time. And when I took it seriously, treated imaginary time the same way I treated imaginary numbers – as clues to a fundamental truth – I began to move along the path that would lead me to understanding. You see, an imaginary number divided by another imaginary number is a real number. Imaginary time divided by imaginary time is real time.

But I jump too far ahead. The truth will be clear soon. You will see this as clearly as I have. And then, I believe, you will begin to understand what I mean.

Yes, I hear the questions you ask. The answers will come to you as you live the moments of your natural life. Maybe you have just been seated in a busy café, perhaps in a city you seldom visit, and you suddenly look up, as if you have been summoned from a dream. You will look around you, at the faces of the other patrons, the child in the corner talking to an old man, the smiling server taking your order, and you will be struck by a vague, persisting familiarity; you will realise that you are in this room, in this space and in this time, with these people, because you are entangled with them. And you will understand that you have come to this realisation because we – you and I – are also entangled.

It will be impossible to describe you, to provide an accurate account of you, of the events that describe you, without also noting that you read this manuscript. Just as it is impossible to describe me, the events known as Ezeani Kobidi, without describing Anyanwu, without describing my mother, without describing my father, my brother, my sister, my wife – and my infant son, Njoku; and without, of course, describing you.

*

I have failed to mention my friend, Ope Adesola. These insights on entanglement did not come to me through rigorous review of the empirical evidence. They came to me fully formed, in a complete epiphany, as I sat on the deck of my home in Ithaca and thought of a red ball and the face of my friend, Ope Adesola, as Anyanwu beckoned him to slit the throat of Begha, a prefect at King's College.

I started crying before I could see what I should be crying about. I felt the need to cry before my eyes produced the corpse that should have triggered my tears. I felt, and then I knew, before I saw. What we know is eclipsed by what we feel.

The Red Ball

The day of what was to be the last full day of our last summer on Martha's Vineyard, I woke up in the late morning and wandered out of our bedroom to see my wife sunning herself on the patio, a large straw hat balanced on her head. I sauntered onto the deck and sat down in the lounge chair beside her.

'How are you, sweet?' she said, then she raised her voice slightly and said, 'Rosa, can we get an omelette out on the patio for the professor, and a glass of orange juice?'

'Where is everyone?' I asked, as I pulled the lever and let the back of the chair recline to an angle that matched the one at which my wife was lounging.

'Philip and Ope went out on the bikes. They wanted to tour the island before they leave. Their flight is at 3.20 p.m. They need to get off the island by one. '

'I am surprised Ope was able to get up so early; I was up with him till very late. We talked for a long time,' I said.

'It must have been quite a conversation. He seemed to be in a good mood,' my wife said.

'He is my best friend, I suppose. Perhaps, he is my only friend,' I said, and turned to look at the weathered planks of the deck.

'You have more friends than that,' my wife said, as she grabbed my left hand in her right.

I looked out at the sand and water in the distance.

'Ope surprised me,' I said. 'He told me that I may have saved his life. At King's College, when he was being bullied and full of spite, he said that I was his friend. He told me that I let him cry beside me on the narrow beds of our dormitory and would push back on some of the seniors pushing him around.'

'Oh, that's so sweet, Ezeani,' my wife said.

'I don't have any memory of any of this, Heidi. When he was speaking it seemed to me that he was either making it all up or confusing me with someone else entirely,' I said. 'It was a surreal experience. He was emotional, quite affected.'

My wife squeezed my hand.

'Before we finally went to bed, Ope rose and walked toward me. I couldn't tell what he wanted. Then he hugged me. It was quite awkward.'

My wife smiled. 'It's sweet. He's your buddy. We should have him over to Ithaca.'

Then the shadow cast by the maid's body fell across my face. I was sitting up, waiting for Rosa to place the omelette on a side table, when I heard the loud crash.

My wife was the first to pass through the front door, making her way quickly up the incline. As I trudged up the hill, I saw Philip from behind her shoulders, his head incongruously fitted in a white helmet, on his knees beside a large white automobile.

A red ball rolled down the road, passing by my wife, and continuing its movement until I stopped it under my foot. There were children gathered at the top of the road, seven of them, beside the ice-cream vendor's stand.

A police car and an ambulance, their lights whirling but the sirens silent, arrived at the top of the road before we got there. Moments later, the ambulance would take away Ope Adesola's dead body, while a policewoman took a statement from the weeping middle-aged woman that had been driving the white vehicle which had just killed him. The policewoman's partner took a statement from the ice-cream vendor, making notes in a pad with a black cover. Philip was sitting on the curb, his head between his hands. He was hyperventilating, his breath drawn sharply into his body in discordant waves.

I stood beside the ice-cream vendor's cart, listening impassively to the conversations around me. I had ascertained the facts. A boy playing with a red ball had run out to retrieve it as it entered onto the road. The woman driving the white automobile had swerved to avoid him and crashed into Ope Adesola as he rode past on a bicycle, pushing him against a light pole and killing him, it seemed, almost instantly. I looked at everything around me. My tears had stopped. I had a small smile on my face and moved my head up and down. My wife grabbed me around the neck and wept. When the policewoman started talking to Philip, I left my wife at the corner and walked back down the slope to the large house we had rented for the summer.

When I got to the front door, it was opened before I could push on it.

Anyanwu stood beside the door. He was wearing his white, black and red cloth around his waist. 'I am not surprised that he was killed, run over like a dog. That boy is a coward,' Anyanwu said, glaring at me from under his floppy cap. 'He is one of those frightened people.'

I jumped as I struck at Anyanwu, throwing my entire weight into the blow. He easily caught my arm. Then he glared at me as he let me down gently to the floor. When a maid's footfalls moved in our direction, Anyanwu stood up and walked out through the open front door.

Trading Options

The August dusk was not cold but there were hints of the cold to come. I was walking home with Rayburn beside me. The conversation was so involved that I did not notice the car keeping time beside us. Not until Rayburn stopped talking and fixed his eyes on the black SUV. Rayburn's lips were pursed, and the beginnings of anxiety started a question on his face.

As I looked past Rayburn at the large vehicle, one of its rear tinted windows was drawn down by a whirling electric motor and my brother's face emerged.

'Chief Ezeani,' he hailed me and smiled. 'You are a tough man to reach these days.' I was surprised to see him.

'Hello, Nnamdi, how are you?' I responded. 'When did you try to reach me?'

'I have tried calling. I have sent letters. Your wife, that Norse ballbuster, is preventing me from talking to you! Can you imagine how messed up you have to be to do a thing like that!'

I made no response. Then I said, 'What are you doing in Ithaca?'

'I came to see you. I am staying at the Statler. Can you believe I got the same suite we stayed in together? Remarkable coincidence,' he said.

I did a quick calculation and concluded it was not a remarkable coincidence but was consistent with what probability would suggest, given the limited number of suites at the hotel. I explained this calculation to my brother. He thought about it for a moment and then said, 'Of course you are correct. On analysis, it isn't particularly remarkable. I suppose it felt that way because I haven't seen you in such a long time.'

We continued this odd conversation, my brother speaking from the open window of a vehicle rolling slowly beside me and Rayburn. Then my brother turned and asked the driver of the SUV to pull over.

When the car stopped, he stepped out, walked up to Rayburn and shook his hand. 'I am Nnamdi, Ezeani's big brother.' He was wearing well-fitting, expensive clothes, his face groomed.

'I am pleased to meet you,' Rayburn said. My brother turned to me.

'The lawsuits have been settled. I have been vindicated. They didn't pay any damages. That is my only regret. It was settled with no admission of liability by either party. Chijioke said that is the best that we could get.'

'Perhaps I should leave you two to speak,' Rayburn said. 'It was a pleasure to meet you, Nnamdi.'

'A pleasure to meet you,' my brother responded. 'My driver can drop you off wherever you need to go.'

Rayburn shook his head to decline and then turned and walked away. My brother and I continued walking, while the SUV slowly stalked us.

'I am setting up my own hedge fund. I've given you a fifteen per cent share in it. That's going to be worth an enormous amount of money,' my brother said.

'What is a hedge fund?'

'It's just a vehicle for managing money. I will be using a

stochastic trading strategy. In a way, it's inspired by your work. Timing is everything in trading, and understanding the true meaning of time has implications for how one can mathematically improve trading performance.'

My brother's statement surprised me. I was intrigued with his suggestion that my work could be used in any such way. It was not an idea that had ever occurred to me. I asked him some questions so I could better understand. After a few of his responses, it was clear to me that, as with his assertion on the likelihood of getting the same suite, he had not thought it through. It also appeared to me that some of the statements were vague and non-committal, as if, at a certain level, he wished to conceal what he was actually saying.

'I heard the wonderful news!' he said. 'Who would ever have believed that you would become a father before me.' My brother Nnamdi smiled, pushed his hand out and shook mine. Then he surprised me by pulling my body into his chest. 'I am so proud of you, Ezeani.'

Then he turned from me and walked back towards the SUV. 'Why don't we have dinner tonight? Please call me at the Statler. I will choose a place. Let's really catch up,' he said as he climbed into the car. When the large black vehicle drove past me, he waved from the window and said: 'Chijioke and I will be here next weekend for the christening, if you can't make it out tonight.'

When the car had driven past, I noticed that I had reached the opening of the driveway for the house on Hanshaw Road I shared with my wife Heidi. I turned and walked up the drive.

That evening, after dinner had been cleared off the table and my wife had given our child to the nanny to put to bed, I told her about the strange interlude.

'He said you prevented him from speaking to me,' I said.

'Of course not. He was calling the house repeatedly, maniacally, insisting on speaking with you. It was disruptive and becoming abusive. I just asked the phone company to stop calls from his number. When he continued calling from strange numbers, our lawyer wrote him a letter and he did stop. Immediately.'

'He has started a hedge fund. He gave me a share of it. He said it should be worth a lot of money.'

'I am glad he is getting back on his feet. But you still need to be careful. He lacks credibility. He has been sending out marketing material to money managers where he lists you as a partner in his fund.'

'Because of the fifteen per cent share he gave me?'

'Well, he has stopped saying you are his partner now. But there is still the implication that you are involved and your ideas somehow underlie the trading strategy.'

'He is my brother,' I said.

'I know, darling,' my wife said. 'I invited him and his girlfriend to the christening.' And she laid a palm on my cheek.

Angel Dust

The day after Ope Adesola died, I was chauffeured back to Ithaca, along with my wife, in a rented van. We did not make any stops on this trip. This was unusual. In previous summers my wife had driven and we had made frequent stops on the road back – to try a meal at an isolated restaurant a few miles off our track, to buy fuel and snacks at a gas station filled with large trucks, or to pause on an embankment cresting a steep hill and survey the lush vegetation spreading to the edge of the sky.

On these trips in previous years, alone with my wife, after the relaxed, expansive days and evenings on Martha's Vineyard, our conversation had been light and dreamy, the easiest conversations of our marriage, as if we were slowly returning to wakefulness after pleasant, diverting slumber.

When we arrived at our home in Ithaca, I climbed up to the porch, my wife's voice dealing with the driver behind me. I mounted the stairs and walked into our bedroom and, still clothed, lay myself down on the bedcover. I fell asleep almost immediately. My time stopped.

The next morning, I rose before my wife, and after I changed into a pair of jeans and a T-shirt, walked out the back door. The sun had just started to rise and there was a sweet smell in the air. I cast my eyes to rocks shelved at the far end of our backyard and saw Anyanwu writing on the ground with a stick. He waved at me tentatively, then continued. He was deeply engrossed in what he was doing. I walked to him.

He seemed startled to sense me by his side. His first instinct was to try to conceal what he had written, but he hadn't moved quickly enough. I was surprised. He had scrawled a sophisticated chemistry formula. 'I am learning science,' he said, and chuckled nervously. As he spoke, Anyanwu was moving the stick across the ground, erasing the chemistry notations.

'Ezeani, it is good you have risen early,' he said when he was done. 'Come with me. I have something to show you. It is something that should have been done a long time ago.' He rose and we walked into the house and up the stairs. Anyanwu led the way. He opened the door of the attic room, and after I walked through, closed it behind us.

'Everything changes. And the world changes with it. This is not how others built their Obi, but this is where your own Upright Lodge will be.' He threw open the white wooden shutter of the small wardrobe built into the wall of the attic room. My Ikenga was mounted on a wooden platform, like the elevated altar of a shrine. The Ikenga's raised right hand held a curved sword, and in its left hand, lower, almost as if it was attempting concealment, was a severed head with a sunburst tattooed on its face. It was the same Ikenga my father had delivered to me years ago, when I was still a graduate student, as we sat in a café at the Statler Hotel.

'Ezeani, here is your Obi. Here is your Ikenga. May nothing accost you from above or impede you from below,' he said. 'Let the kite perch and let the eagle perch. It is the one that says the other shall not perch who invites disaster.'

Anyanwu was standing with his fists against his hips and spoke proudly, his chest starting to perspire. As he opened his mouth to speak again, my wife's voice came to us from the stairs: 'Ezeani, are you in the attic?'

Anyanwu laughed. 'Go and answer her. Come back later. I will have something for you.'

'There you are,' my wife said, as I came down the stairs. 'I thought we might go out for breakfast. Let's go somewhere nice.' She slipped her arm into mine.

A week after our return to Ithaca, in the small library within Clark Hall, I opened a chemistry textbook and searched for the notations I had seen Anyanwu scrawling. The formula described phenylcyclohexyl piperidine, a white crystalline powder easily dissolved in water or alcohol. In another book, I learnt it was a dissociative anaesthetic with significant mind-altering effects, including hallucinations, distorted perceptions of sounds, and violent behaviour. I also learnt that the compound was commonly referred to as PCP or Angel Dust.

15

THE COPENHAGEN
INTERPRETATION

On the Road

After our last summer at Martha's Vineyard, after the slow weeks that marked the beginning of fall, the days started to pass quickly. I was happy to immerse myself in work. In the morning, my wife would make breakfast and we would sit, side by side, and eat. Then she would drive me to work in the physics faculty building. My days were occupied in a blur of exertion.

In the evenings, I would usually walk back, through the leafy campus, to our home on Hanshaw Road. I was often accompanied by one of my colleagues, usually the professor or graduate student who I happened to be talking to when the urge to return to my home overwhelmed me, and I abruptly turned and headed for the door.

Surprisingly, it was in this period that I mustered a sudden enthusiasm for the lecture tours I had only grudgingly accepted when my wife had initiated them just two and a half years before. So, it was in the late fall of that year Ope Adesola died that we expanded the tour further down the east coast of the United States, all the way to Florida State University. It was an institution only notable, as far as I was aware, for the fact that the great theoretical physicist Paul Dirac had chosen to retire there after his years at Cambridge. For this trip, my wife made arrangements for us to travel on the Amtrak sleeper, in a private car. As the train rattled on the track, I stared through the windows and then thumbed through my old copy of Dirac's *The Principles of Quantum Mechanics*.

My wife sat across from me facing the space the train had just passed. She was sipping red wine in a clear glass.

'Tell me about him?' she said, as she pointed at the book with her glass.

'He always followed the mathematics. He really shaped what physics means to me. When you take the square root of any number, you always get two answers. One is positive and the other is negative. Before him, everyone just ignored the negative answer, assuming it must just be nonsense. He didn't. He assumed it was saying something valid about the universe and that was how he theorised antimatter.'

My wife smiled. 'He peeked in Chi's Upright Lodge.'

'Well, actually, he didn't. He just computed it must be there, that the negative answer must mean something. That's the thing about theoretical physicists. Technically, we don't discover anything. We just point to where others can find things.'

'Like map makers,' my wife said.

'Yes, like map makers,' I said.

The train jolted and she instinctively pushed her hand away from her body so that the wine in her glass, if it spilled, would be further from her body and less likely to stain her.

The visit to Tallahassee was a delight. A dedicated postdoctoral researcher, who my wife informed me had eagerly appointed herself our guide, was at the train station to pick us up. She walked up to us as we loitered at the station's entrance, feeling the heat rising from the concrete pavement. She was short, her hair in locks, and had smooth, reddish-brown skin. Two beads of perspiration walked across her brow.

'I am so honoured to meet you, Professor Kobidi!' she exclaimed, reaching her hand out to me. 'My name is Ngozi.' As I completed the handshake she turned to my wife and hugged her. 'It is so wonderful to meet you, Mrs Kobidi. I feel I know you already from all the emails.'

My wife smiled. 'Please, call me Heidi. It's nice of you to fetch us, Ngozi.'

The postdoc walked us to the parking lot and opened the doors of her small car. 'I can't believe this is happening,' she said once we were all seated, my wife in the front seat beside her. 'Sorry, the air-conditioning isn't working.' She pressed toggles that brought down the car's half-raised windows.

Before she conveyed us to the hotel, she made a detour to the university campus. The small car pulled into a parking lot. 'Unfortunately, we need to walk the rest of the way. I thought you would want to see this first.'

She showed us around the Paul Dirac Science Library and then insisted that I and my wife sit on the bench beside the Dirac statue on the front lawn, so she could take a picture. When that was done, she requested that we take one with all three of us in the frame, soliciting the assistance of a passing student. Then, she insisted, 'Just the great physicists,' and asked me to sit alone on the bench with Dirac's metal sculpture. I shuffled close, and placed my hand on its metal knee and she took several photographs, looking down at the camera and smiling.

That evening, I enjoyed my talk and the air of fuss that hung around me in the cool, large lecture room. The next day, while my wife slept in, the eager postdoc took me on a prearranged visit to Florida State University Libraries' Special Collections and Archives division. As we walked from her car, she turned to me and said: 'I cannot begin to tell you how much it means for me to meet you. You inspired me to pursue a PhD.'

'What do you work on?' I asked.

'Condensed matter physics,' she said.

We continued walking in silence, the postdoc slightly ahead of me. Then she stopped on the pavement and turned around to me. 'I don't know if it's an appropriate question for me to ask, but in your talks you mention Anyanwu a lot. Does he actually appear to you?'

'Yes, he does. As clearly as you appear to me now,' I said.

She stared at me for what seemed like a long time. I sensed words forming in her mind. She did not speak. I could not tell what she was thinking. Then she looked down at the pavement and started to speak, in a low, purposeful tone. 'The thing is, I just wonder, if it's just a device? I wonder if it's a way for you to introduce your Igbo roots into the conversation. My parents are from Nigeria as well and I think I get it.'

I looked at her and then smiled. 'I don't have a choice. I have to talk about him. Anyanwu will not let me do otherwise. He is coercing or browbeating me,' I said. 'I am not certain which term is more appropriate.'

She fastened her eyes on me. 'Do you really see him?'

I nodded my head.

'Like, is he really there? Do you see him?' Her voice had raised a few decibels. I continued to stare at her. She seemed suddenly embarrassed. 'I'm a diagnosed schizophrenic. I've had a mental breakdown,' she blurted. There was pain in her eyes. I sensed she wanted me to say something more. I did not know what it was. Then she said: 'I am sorry. I know this is inappropriate.'

We continued to walk in silence.

We soon arrived at the top of the library stairs where a tall, balding archivist stood waiting for us. After an introduction in which Ngozi referred to me as a 'legendary physicist', the archivist took me into a dark room and left me for six hours and thirty-five minutes with the university's vast collection of Dirac's papers. When I walked out Ngozi was waiting in the lobby. She drove me back to the hotel. She did not speak in the car.

As we walked through the hotel parking lot, I said: 'We are the lucky ones, people like you and me.'

'How?' she asked.

'We are fortunate because we realise that even our existence as distinct persons is an unchallenged assumption. Most people thoughtlessly assume they know who they are. We don't assume this. We know we don't know who we are. We have a superior awareness of the actual truth. Not like our deluded brethren,' I said.

'It doesn't feel much like fortune,' she said abruptly, her face turned to the pavement. 'I have tried to kill myself.'

'How could you have failed?' I asked, stopping and staring at her. She stopped too, but her eyes remained on the asphalt. After a moment passed, and she had not responded, I said: 'Perhaps because your life is precious to you; even when you don't know who you are.' I turned my eyes down to the pavement.

'What happens when it stops being precious to me?' she said.

I looked up.

'I don't know,' I responded.

The Halting Problem

One morning in the week after we returned from Florida, as

Heidi was driving me to work, she made the decision to stop at a pharmacy. As she parked the car in the lot, she said: 'Just a second, Ezeani. I'll be right back. Need to pick up another pack of the follicle-stimulating hormones. I will be gone for just a second.' She stepped out of the car. As she strode into the building, she turned the key fob back at the car. Two small beeps reassured me the car was locked. I sat in the passenger seat looking at the skyline through the windscreen. I must have become so preoccupied with my thoughts that I lost my sense of self. It was as if my time had stopped. I was rattled back to the car and the parking lot by a sharp knock at my side window. As I looked out I saw a short man, in a large dirty coat that hung down to his ankles. The man was pushing a shopping cart, laden with boxes and various electronic items. I turned my face from the man and looked out of the windscreen, trying to avoid contact with his eyes. The dirty man rapped several times, sharply, against the window. There was aggression and anger in the displaced beats of the man's rapid knocks. I turned again to the side window. He had brought out his right hand, covered by a multicoloured woven mitten, and beckoned me to bring down the window. I felt a small fright. I reached into the cup holder and picked up some of the money my wife habitually left in there and, carefully cracking the window open about an inch, threw it out to the small, dirty man. I could see his face turn towards the ground. He stooped and picked something up. I presumed it was the money. Then he started banging sharply on the window. The man seemed incensed. I was frightened. He was now throwing punches at the window. Then he bared his teeth at me. The face looked familiar, but I could not place it. It was just as a sound startled the man and he turned his face away from me in profile that I saw there was a gash and blood on the left side of his head. It was in just the place where

Ope's skull had been crushed. The blood was plastered against his skin in the same way. Then my nose filled with a stagnant, metallic odour that started to make me nauseous.

The man threw his face against the window. It was Ope! He was ranting. I could not hear what he was saying but his finger was pointing accusingly at me. I listened to the voice. It was not Ope's. The voice was shrill and loud. Then I looked at the man again. It was Ope!

Quickly, I turned again to the windscreen and fixed my gaze. I ignored the man at the window. I started to nod my head. I was afraid. Then the man stopped. Hunched over, he pushed his cart towards a large red car that had just pulled into the lot. I was relieved. As the man walked away, I began to doubt it was Ope. It was the gait which made me doubt. The steps were the slow shuffle of a defeated spirit. That was not the way Ope walked.

When my wife returned to the car, clutching a brown paper bag, I told her what had just happened. She listened to me, her face expressionless. Her right hand was resting on my left thigh. 'I am afraid the trauma of Ope's death is taking its toll. Have you discussed it with Dr Chan? It is important that we don't let anything undermine the amazing progress you've made.'

She started the engine. I was surprised she made no attempt to verify the identity of the small dirty man still standing beside the red car. Just as she was about to place the car in reverse, Heidi stopped her motion and turned to me. 'Have you ever had any hallucinations like this before? Other than your conversations with Anyanwu, have you had any conversation like this?' she asked.

'No,' I responded. I was not lying. I had just forgotten the moment in childhood when a lizard that was about to die had started to speak to me through lips protruding from a battered skull.

Wave Collapse

Reality is not what you think it is. At the core of reality, nothing is tangible. All that exists – if that is the right word – is the spirit-sea of probability waves; the quantum waves of potentiality. When I was still young, Anyanwu had shown me the lingering world of possibility that emanated from, and was mirrored in, the existence of Chi Ukwu, the Great Chi, and reproduced in the Chi of everything in the universe. I had seen that the universe was suffused with this bountiful potential and its dancing, uncertain manifestation.

At the core of our reality are probability waves. A wave is a propagating dynamic disturbance. This is the summary of everything. At the core of our universe, at its most fundamental level, there are waves. Stop, and keep that in your mind. Let that settle with you.

Feel your Chi; sense your possibilities, the full extent of what you can be. You may close your eyes if it helps. Sense what be-ing you means; feel your Chi and how large it spreads, the waves reaching out to the universe, to all space and to all time. When you have wallowed in the spirit-sea of these probability waves, when you have observed how what you are and what you can be are intimately related but yet utterly apart, you will also sense the tender bond between Chi and Eke.

And when you have experienced this joyous bond, between what can be and what is, the persistent interplay that makes the universe, and of course, you and I, you will begin to sense my repulsion for the concept taught to most postgraduate physics students: the abhorrent idea that these probability waves collapse to take on a singular aspect; the notion that what we can be is ever reduced to what we are. This repugnant theory propagated like dogma by men with – and I am afraid there is no other characterisation that fits – fascist leanings, as the edict of their Copenhagen institute, is an approach that I instinctively rejected as a student of physics. I ignored their postulates but never made a show of this disengagement in my early papers. It was not necessary at that point to make my dissension clear. I could lay out my initial ideas without challenging their noxious orthodoxy.

But when I published my last theorem, when I made clear my ideas were inconsistent with the absurd, deceptive high-handedness of their mantra, it was the adherents of this school, led by my ostensible colleagues Hess and Huslander, who coordinated the attack against my academic reputation. They led the assault against my last theorem, the most important work of my life. The theorem that their fellow traveller, the unqualified Dr Chan, PhD, dismissed as my 'delusional hypothesis'.

In the letters I sent to most of the nation's key physicists, I addressed those, like Philip Bousquet, who objected to my characterisation of these individuals as fascists and Belgians:

What term but *fascist* should I use to describe physicists that openly state: 'The tenets of quantum theory are long established. It's a closed theory and the basic outlines and principles are not subject to change or modification'?

What word but *Belgian* adequately describes characters whose wickedness is so depraved that it manifests, takes a body and wears a hat, just like their much-mourned King Leopold? This is the character of these men, and that of their cohorts, who conspired to attack me.

As they attacked my theorem and reputation, they were assisted in their task by the assaults on my stability, confidence and peace launched by my wife Heidi. An assault intensified by the efforts to separate me from my son Njoku. This was a tremendous burden on me, but it would not by itself have been successful in silencing me they had not, to my utter surprise, had the help of the unreliable Sun God Anyanwu. It was he, Anyanwu, who ultimately succeeded in closing my mouth.

It was only at the end that it became clear to me. It was not Anyanwu's exhortations that had led me to acknowledge the Sun God to the world, but my wife Heidi's appeals to pragmatism, and of course my vanity. The increase in Anyanwu's power and the diminution of my existence are linked. As he had betrayed Ani and the other Gods of their firmament, Anyanwu would also betray me. They were in the conspiracy together. Heidi and Anyanwu conspired with others to enslave me, to sell off my ideas and to denigrate my value. They plotted to kill me off, and with me, my ideas. As far-fetched as this might appear to you, it is the truth.

I only realised this at the end. I was utterly deceived.

Baptism of Fire

The parish priest of the Church of Immaculate Conception visited us at our house on Hanshaw Road. He arrived on a windy August Saturday afternoon dragging a yellow-brown leaf under his left shoe. I was seated in an armchair in the living room. As he walked in, he looked briefly at me and said: 'Good afternoon, Professor Kobidi.' I did not respond. He sat down in the other armchair. I sat still and stared at him. He looked away from me. He seemed uncomfortable, tapping his left foot repeatedly against the heavy carpet.

We sat in silence until Heidi swept into the room and said: 'So kind of you to come, Father O'Malley.'

When Heidi stopped and glanced at his restless foot, the parish priest looked down and noticed the leaf. He bent over and lifted the dead leaf, crumpled it and then turned to scan the furniture for a place to drop the refuse. Before Heidi could offer a solution, he quickly stuffed the leaf into a pocket within his black cassock.

'Have you decided on a name for the christening?' the priest asked.

'Yes,' Heidi responded. 'Our son will be called Njoku.' She walked over and perched on the arm of the chair in which I sat.

The parish priest looked uncertain. 'This is a name from the professor's tribe?'

'My husband is Igbo. That's his ethnic group. Yes, it's an Igbo name.'

'That is quite fine, Mrs Kobidi. What will his Christian name be?'

'Well, I am not sure what his other names will be. I would also like the child to have a name from my side of the family. My father insists he will be helping in that choice' – and here my wife smiled – 'but that name hasn't been decided.'

I rose stiffly, turned out of the room and started to walk up the stairs.

'Excuse me,' my wife said to the priest. She caught up with me on the first landing. 'Are you OK? Ezeani?' She lifted my chin and looked at me. There was concern in her face. 'I know it's a lot. The child, the lectures, everything,' she continued. 'Please believe me, it will get better. We will get through this.'

I looked at my wife and tried a smile. Then I walked up to the attic room in which I had started taking naps in the months after my son came home and the house's other floors became consumed with the noises of the infant and his caregivers.

When I opened the door, Anyanwu was lying on the bed, his boots on the bedsheets. 'What did that priest come here to do?' he asked.

On the Saturday morning set for our child's baptism, I woke to find my wife hovering over our bed with a phone held against her ear. At some point she raised her hand and covered her mouth. When she hung up there was pain in her face.

'That was Father O'Malley. Someone tried to burn down the church.'

'What time is it?' I asked.

'We won't be able to perform the ceremony at the church. It's terrible.'

I was staring at my wife. She continued to speak.

'We've invited quite a number of people. Father O'Malley has offered to come over and perform the rites in the house. What do you think?'

'What time is it?'

'Maybe 7 a.m. What do you think?' She looked at me. 'We were going to have people over after the ceremony anyway. We can just have everything here.' Then she paused and said: 'It is awful.'

'What caused the fire?' I asked.

'They are fairly sure it's an arsonist. Apparently, the entire church had been doused in some sort of flammable fluid, but fortunately only the sanctuary was successfully set aflame. Nevertheless, there is extensive water damage from the efforts of the fire department.'

'Why would anyone want to set the church on fire?' I asked.

'The world is full of crazies,' my wife said.

The fire would make it into the newspapers. Several months later there would be an appeal for funds to renovate the damaged church. My wife's family wrote a contribution equal to a small fortune, single-handedly paying for the repairs. This was all covered in the newspapers. I was surprised by the open, public generosity. When I commented on this to my wife, her reply was short and sharp: 'We did what we needed to do.'

Phase transitions, familiar in everyday occurrences such as the freezing of water to ice, creates strange and surprising effects when the bodies involved are at temperatures approaching absolute zero. These so-called quantum phase transitions generate distinct coexisting states of matter, resulting in bizarre collective behaviour.

It was at the moment of a quantum phase transition, when the heat between us was approaching zero, that I discovered my wife – and her father – in fact suspected me of involvement in the arson at the Church of Immaculate Conception.

Living with Jack

My son reached his small hand out to me. He was tottering and his steps quickened, in an awkward gait, as he sought to keep himself upright. He could sense he was about to fall and was making quick adjustments to maintain balance and his forward momentum. I reached out my hands and caught him. I held him up. He bubbled with pleasure, a wide smile spreading on his face. I picked him up and placed him on my knee. Then I kissed his forehead. His body radiated pleasure and he gurgled, making speech-like noises. Then my son reached out his right hand, touched my nose and slid his fingers between my lips.

I smiled.

My wife who, with her shoulder leaned against the door jamb, had been watching me and our son, moved into the room and took a photograph of Njoku on my knee. In the photograph we are both smiling broadly. And although he was still an infant and I was at that point a middle-aged man, the resemblance between us is uncanny.

Phase transitions are an important theoretical and practical aspect of the physics of our universe. In a phase transition, a small change leads to an almost magical transformation in certain qualities of a system. Add just one degree of heat to water at ninety-nine degrees and it transitions from liquid to wisps of gas.

It is a strange fact of my life that I had minimal interaction with, and only mild affection for, my child until he started to walk. Then, as if a body experiencing the most remarkable phase transformation, my senses were overwhelmed with affection, care, concern and anxiety for his welfare. This phase transition was one I had previously noted in literature but not truly understood. Instantaneously, I had fallen in love with my own child.

16

THE SPEED OF CAUSALITY (OR THE PERMEABILITY AND PERMITTIVITY OF SPACE)

The First Internment

The fiery circle circumscribed the blistering darkness of the black hole floating above my hospital bed, mere inches above my head. The fire burnt with a piercing howl, like an angry gas turbine. The heat singed my face. I lay still, wide-eyed, staring at the burning orb. I did not know what I was looking at. And I was filled with dread. It was only when another black hole appeared, to the left of the first, circled by another roaring disk of fire, that I started to perceive that the black holes and their flaming boundaries were actually three-dimensional. Once I realised this, it took me only a few moments before I recognised that the burning spheroids were the angry eyes of the Sun God Anyanwu. The rest of the room was pitched in blackness. My fingers and toes were numb. I felt cold.

I started to scream at the top of my voice, pulling my arms against the restraints that held my wrists to the sides of the bed. The sound that came out of my mouth was a howl, an unending shriek, sustained by simple terror.

'Shut up!' Anyanwu bellowed as he moved his head even

closer to my face. 'I am the only one protecting you!' His eyes burnt my flesh.

'What do you want?' I pleaded. I was frightened.

He hissed, rose from the bed, and walked into the darkness. I could see the red and orange flames of his eyes in the pitch black at the furthest corner of the hospital room.

'Tell them I am your God!' he shouted, as he suddenly ran straight at me, like a charging animal. In panic and terror I closed my eyes. The sound of the fiery furnace lingered in my ears and then subsided. The heat warmed my face. I kept my eyes closed. Gradually, consciousness drifted away. My time stopped.

When I opened my eyes again the hospital room was filled with white, fluorescent light. There was a male attendant in a white tunic standing over me. He was making adjustments to my restraints. Anyanwu was squatting in the corner; the red, white and black checkered cloth was tied around his waist, and a blue disposable surgical cap covered his head. The hospital lights hummed.

'Have you woken?' Anyanwu said, looking at me. Then he looked down at the grey floor.

'Where am I?' I asked.

'The same place you were yesterday. Hutchings Psychiatric Hospital,' the male nurse said.

'A place where they keep mad people,' Anyanwu muttered from the corner, and then spat on the floor.

'You can't keep doing that,' the male attendant said and looked pointedly at me. 'Someone has to clean it up.' And then his nails bit into the flesh of my wrist. I shrieked in pain and looked at him.

Then I screamed again in terror. I recognised him. It was Jesus. He smiled. And then he turned and walked out of the room, stopping to glance at Anyanwu, who stood up, placed his hands on his hips and stared at the man in the white tunic. Jesus dropped his gaze as he passed through the door.

As the door of the hospital room closed behind Jesus, Anyanwu leaned his body against the frame as if he was blocking it so no one could enter. His eyes were fixed on me. They were no longer rimmed with fire. They looked like normal eyes. We stared at each other.

'Are you ashamed of me?' he asked sharply. He was watching me, his eyes following the movement of my head. 'Don't lie to me,' he said quietly, gently. His voice had changed. It was the sound of my father's voice. 'Don't be ashamed of me.'

I was kept in restraints for just the first two weeks of my first involuntary confinement. By the end, I was not only unrestrained in my room but was allowed to walk the halls and the grounds of the institution freely. At the end of this period, my interviews with the white-coated psychiatrists no longer consisted solely of long visits to my room. I was permitted to call on them at their offices located in a separate wing of the large facility. In those interviews, I would be invited to sit on a couch or armchair and engage in stilted conversations. I was always sceptical of the scientific basis of their medical claims, and in the beginning I engaged them often in extended conversation on these points.

'I understand that you are under the impression that I am mad and you are sane, Dr Bellow, but I am curious how you are able to dismiss the hypothesis that things are the other way around; that I am sane and you are mad?'

'It is not an easy hypothesis to dismiss. But this hospital here, the doctors, the establishment: it is some proof that my view is also shared by many others. And my view is that you are currently suffering from a mental illness, and like any other illness it hinders and harms you. I am dedicated to making you better.'

I smiled. 'Your only response is an appeal to numbers?'

Dr Bellow looked at me, a small smile on his face. He was always patient. He spoke only when he had to. This discipline initially led me to make the mistake of filling the silence with words. Words I would later understand were used against me.

'Do you know how many people used to believe the world was flat?' I clapped my hands in triumph, gleefully chortling. 'Almost everyone. The world is full of lunatics. They are in the majority. Everywhere you go.'

Dr Bellow did not speak immediately. He was holding his head still, eyes on the notebook on his thigh. And then he looked up at me and asked: 'Why do you think you are different?'

I responded immediately, my words filling up the silence. 'I am not different. That is my entire point. I am not different at all. Neither are you. We are the same. We are just constructs, you see.'

'What do you mean?' Dr Bellow asked.

'We are made of atoms. You and I. Everyone. But atoms are not made of things; they are representations of the interactions of probability waves; conversations between an observer and the observed; between the thing that stands and the thing that stands beside it. And since all this is true, as fraudulent and absurd as it appears, what do we know of our life? What can you say "you" are? What can it possibly mean if we are all just these manifested ripples of energy jumping in an infinite field of probability, both observer and observed? And just before you get carried away and start to tell me about the subjective experience of life, Doctor, pay attention to the fact that it is these same probability waves whose interactions create these apparently subjective feelings of joy, pain and beauty!'

I slammed my palm hard on my thigh in triumph. It was only after I glanced at Dr Bellow that it occurred to me that he might not have understood how thoroughly I had destroyed the smug point he had been making. He started writing in his notebook.

Sound Advice

In the third week of my first involuntary confinement, I was ushered by a large orderly into a beige room with giant windows and anodyne paintings on the wall. My father and brother sat side by side on a couch. Chijioke sat by herself in a chair close to the wall.

Permeability is the tendency of a substance to allow another to pass through. Eventually, water seeps through rock. Eventually, the advice I was being offered sank in. The documents were produced and I signed. Chijioke brought a retractable stamp from her briefcase and banged it below my signature, then scribbled a mark asserting she had witnessed my hand.

I understood from the things that my brother said during the visit that my father had insisted on visiting me; that my brother had agreed to pay for the trip for my father and his wife, Maria; that my father would next visit my wife and son; that he would then be flying to Florida for a vacation before he and his wife returned to Nigeria.

My brother held my face between his hands. He hugged me again, more tightly and longer than he had when I had first entered the room. Chijioke gave me a brief hug. My father shook my hand. Then they walked through the door.

As the weeks of my first confinement ran on, Anyanwu started to accompany me on my walks down the halls and around the grounds of the large complex. One late morning, as we passed through the lawn at the back service entrance, we saw Jesus, still dressed in the white orderly's outfit, speaking to a female nurse. She was dressed in purple scrubs. They appeared to be in deep, animated conversation. As we approached, she stood tall and pulled Jesus to her bosom in a deep hug. 'You are a miracle worker,' she said. As she walked past me, I noticed that she had tears in her eyes.

'Immaculate Conception!' Anyanwu called out to him in a friendly tone. 'You won't leave these fallen women alone.'

Jesus waved a dismissive arm and smiled slightly. As we came close, he greeted Anyanwu with a slap of the palms, like an inverted high-five. 'You know you are a pain,' he said to Anyanwu. 'You can be a real bastard.'

'Now, now, let's not start making references to doubtful paternity,' Anyanwu said, then chuckled.

Jesus didn't seem to have been paying attention to Anyanwu's words. He took out a pack of cigarettes from his trouser pocket and offered one to Anyanwu. 'I really shouldn't be giving you these,' he said.

'I shouldn't really be accepting it,' Anyanwu said, laughing. 'Let me guess. It's poison.' For some reason this comment caused Jesus and Anyanwu to start laughing uproariously, so hard they each independently bent down to hold their knees, the cigarette sticks carefully balanced between their fingers.

When Jesus had straightened up, he turned to me and spoke directly. 'I know who you are. They say you are a big-deal physicist. To hear them talk, it sounds like you're the next Einstein or something.' He smiled. I couldn't tell if he was mocking me. I glanced at Anyanwu. His face was impassive.

'Look, I'd like to talk to you some more,' Jesus said. 'We aren't enemies. Maybe we can even be friends. Maybe, I can help you get out of this madhouse.' Then he stubbed out his cigarette, bumped fists with Anyanwu and walked back into the hospital.

After Jesus had passed through the door, Anyanwu slapped me on the shoulder. 'He is right. We have been talking. He agrees with me. You can get out of here if you stop telling them the truth.'

'I thought you hated that guy,' I said.

'Yeah, I don't like him, but we can work together. He is not a fool. He also sees that he is in danger. Most of his worshippers now are our people; the believers he stole from me. Very few of his people go to his priests.'

'Why would he want to help me?'

'You brought me back to life. There are many people now that believe in me because of you. You don't follow these things, so you don't know. Many people now know of me because of you.'

'I still don't see why you are working with him?' I said.

'It's complicated, but don't worry about it. Just focus on getting out of this place so you can continue your science,' he said. 'Leave it to me to handle Jesus.'

I sat down on the grass and stretched my feet straight out before me. Anyanwu sat down too. He adjusted the blue surgeon's cap on his head. 'My son,' he said, 'don't be afraid. Don't be ashamed. You are part of a great legacy. Do not let them mislead you.' Anyanwu picked up two rocks from the grass and banged them together. Then he discarded them on the lawn.

I sat on that lawn and stared at the rocks for a very long

time. The sense in Anyanwu's suggestion seeped in, permeating my understanding.

Because I have studied physics, I know that the permeability of every substance can be measured, including the permeability of empty space. And the relationship in empty space between permeability, the willingness of one substance to let another through, and permittivity, the determination of the other to resist, is a constant. And that constant is the speed of light. It is not just the fastest speed anything can travel; it is the fastest speed at which one thing can cause another. The speed of light is approximately three hundred thousand kilometres a second. It is fast, but it is not, by any scale, infinite. Things just take time.

I thought about this for several hours and I started to realise its implications for my last theorem. If our connection – our entanglements – is what creates the space in which we exist, then time is the count of these interactions. And there is a limit to how fast we can count. It just takes time.

The Blessing of a Sister

My sister bounced Njoku on her knees. He was holding on to her thumb and gurgling laughter. I was in the armchair staring at them, the morning light flowing into the living room of our home on Hanshaw Road. My wife sat on the chair beside my sister. They were in conversation, in low tones that seemed designed to exclude me. I turned to the window, as if in disinterest.

'I think he is much better now. It's been a horrible experience,' my wife said.

'I can imagine it's been difficult, Heidi,' my sister said. It seemed as if overnight her face had taken the shape of our mother's.

'I have been able to manage. It hasn't been easy. And your brother's insinuations have been very hard to bear.'

'I can't believe Ezeani would say anything bad about you,' my sister responded sharply.

'Not Ezeani. Nnamdi,' Heidi said.

'I disagree with Nnamdi's approach, Heidi; but I must tell you that I do question whether having him institutionalised was necessary. Look at him. It's done a lot of harm. His academic work has been such an important thing in his life. He now has to live this down with his colleagues . . .'

'You didn't see him before he went in. We really tried to keep this quiet, but Nnamdi's lawsuit generated publicity. I mean . . . technically it was Ezeani who sued, but it was your brother Nnamdi's idea. His girlfriend filed the case.'

'I just want Ezeani to be able to live his life. He is vulnerable. And I see what you've done for him. I don't know that he could have had such a rich life – even his career – if you hadn't come into it.'

'Oh, dear!' my wife said, rising from her chair and hugging my sister around the shoulders. My sister rose from the chair, handed my son to my wife and then hugged them both to her chest. Then she turned to me. 'Ezeani, come. Let's go.'

I rose and walked with my sister to the rental car she had parked in our driveway.

'What kind of food would you like?' she said as she opened the car door. I did not speak. I could not think of what I wanted to eat.

My sister smiled. 'Get into the car,' she said. 'We can decide when we are on the road.'

As the car turned to the right at the end of our driveway,

my sister gave off a small yelp. Then she said: 'Free at last!' I looked over at her. 'I know it's unkind, but I really feel stifled in your house. I love the stability and love your wife provides, but I don't like it. Does that make sense?'

I watched her as she raced the car down the street, making aggressive turns at corners. 'That should do,' she said as she abruptly turned into a parking lot, screeching the tyres.

When we were seated in the Mexican restaurant, and the waiter had taken our order, my sister squeezed into the booth beside me.

'Ezeani, how are you?' She held my hand. 'Are you happy?'

'I don't know,' I responded. 'Sometimes I am afraid. I don't know why.'

She smiled weakly at me. 'You have a strong Chi,' she said.

'How do you know?' I asked, and I started to smile.

'I have a strong Chi too,' my sister said, and smiled at me. We started to laugh.

After we had eaten, my sister asked me to come with her, the very next day, to her home in Connecticut where I would spend a few weeks away from my home on Hanshaw Road and all that was bearing down on me. This was barely two days after my release from the first confinement at the Richard H. Hutchings Psychiatric Hospital.

'Think of it as a little relaxation break, away from everything. My son is spending the holidays with his father. It would be a good time for me, and I hope for you. You can rest a little and see things from a different perspective.' My sister stopped talking and looked at me. 'Good idea?' I looked at her. Her face was full of love, the face that my mother had. I closed my eyes and I reached out and hugged her. She patted my back.

The next day, after a peculiar negotiation between my wife

and my sister – my sister wanted three weeks, my wife one – I left for a two-week stay with my sister at her home in Connecticut.

A Sublime Theory

I returned from the visit with my sister revitalised. In the two weeks I spent at her house we barely spoke more than a few sentences a day. I felt I was in a new home, and I was released, as if magically, of anxiety. I read casually from the stacks of books lying on bookshelves, nightstands and the floor of her house. Sometimes she would encourage me to talk about my son. 'Njoku reminds me so much of you when you were a child. He is so bossy and sure of himself,' Obiageli said. I smiled.

'Ezeani, don't forget that. Everything else you are, you are also that boy. The boy that is just like Njoku. Don't forget that you are that, too.'

My sister drove me up to Ithaca after two weeks as she had promised. Also, as she had promised, she came to visit me every month afterwards and called me on the telephone every week. Sometimes I had nothing to say to her. But on certain other days I filled an hour with stories of my son Njoku's remarkable conduct, and sometimes the stories ran so long that it was my sister who would gently insist she no longer had the time to continue the call.

In the period after my return, I dedicated myself to what had before then been nascent, preliminary work on entanglement. The work that, from the very first epiphany, sitting bare chested on the porch of the house on Hanshaw Road, thinking of the connections to Ope Adesola, I had been convinced was the most important work of my life. I gave up the lecture tour and numerous collaborations on incremental work with other physicists. The loss of these activities also represented the loss of certain connections to my wife. She became less integral to my work in physics, and my work by its very nature became more insular. This was a period where we spent very little time together and she restarted the practice of travelling to Italy in the summer with her parents and our son.

I did not take vacations in the summer months. I was focused. Entanglement permeated my being. I worked arduously, prodigiously, on this theory, mapped out the underlying mathematics which, if it did not yet prove the 'delusional hypothesis', certainly set it on a solid, spectacular foundation. It was three years of solid, almost brutal work. One September night, in a frenzy, I compiled its essence into an eight-page paper that I emailed to my friend and colleague Philip Bousquet at Yale. I woke up the following morning to the smell of fried eggs and sausages from our dining room, and an announcement that Philip Bousquet was downstairs waiting for me.

Philip did not want to have breakfast. He was eager that I set out with him to the faculty building. He had contacted Alcott, Rayburn and Scanlan and they were waiting for us at a small seminar room in Clark Hall.

As we walked into the warm room, I saw that the Kobidi Entanglement Equation had already been scrawled on the blackboard and copies of the document I had sent to Philip were scattered on the desks.

'There is no preamble required here,' Philip said. 'If Ezeani is right, this will be the most impactful theory since Einstein's General Relativity.'

Rayburn laughed. 'That's an understatement,' he said.

'Walk us through it,' Alcott said. I pulled at the sleeves of my sweater and began to speak.

The Second Internment

Nine months after that glorious morning, I was taken in an ambulance from our home in Cornell to the Richard H. Hutchings Psychiatric Hospital in Syracuse, New York. It was a warm afternoon, and in the ambulance I made several requests that the air-conditioning be adjusted to compensate for the heat. These requests were ignored.

This was the start of the second period of confinement in the mental hospital, again on the petition of my wife Heidi. I was released from this incarceration almost two months later when Dr Bellow attested that my temporary infirmity had been treated and I was sufficiently cured of my illness, so as not to pose a danger to myself or others. For what was left of my natural life, I would not again see a psychiatrist or have my freedom curtailed in this dramatic way.

I understood why I had been brought back. I had gotten carried away. The success in completing the equations of my entanglement theory and the initial enthusiasm with which it was received in our community of physicists were a great boost to me. There was certainly some doubt about whether all the elements of my theory would be established in the fullness of time, especially in connection with its more radical predictions on entanglement entropy, but it is this way with all truly revolutionary theories. One must always remember it took several years, sometimes decades, for the many implications of Einstein's General Relativity to be proven.

This warm reception and the air of primed promise that followed the publication of my paper in the journal *Nature* were reflected in flattering correspondence, and in a positive commentary published by Phil Bousquet in the *Reviews of Modern Physics*. Unfortunately, unexpectedly, things turned in an unpleasant direction when Tristan Huslander, professor at the California Institute of Technology's Department of Physics and a member of the National Academy of Sciences, uttered scandalous rubbish about me in the ending moments of an interview in *Time* magazine. The stinging quote from that article, repeated in several other publications, was this: 'Kobidi's early work is promising, but this descent into voodoo physics illustrates what has gone wrong with the science. I simply do not take it seriously; why would I?'

And his response to the sensible follow-on question of his interlocutor: 'Yes, it is new, and to be frank, I have not attempted to review his underlying mathematics. Why should I exert the effort?'

That was the first blow. It may perhaps have ended there had I acted on Heidi's advice and let Philip author a measured rebuttal of Huslander's nonsense. I concede that my reaction, several pages long, scrawled in my hand, photocopied, stapled and then mailed to almost every individual listed in the faculty directory of every reputable physics department in the United States, did not help. In justification, I can only say this: I did not feel I should concede my right to defend myself. Not now, when I was no longer a child being beaten in a schoolyard.

As I hovered over him, the clerk at the Kinko's copy store tried to reassure me: 'Don't worry, sir, it will all go out tonight.'

Other blows would fall. Hess published an article in *Physics Today*, where he claimed a willingness to undertake a review of the underlying mathematics and physics from which Huslander had exempted himself. But, without actually making such a review, he assured his readers of his certainty that I was in error, based on the 'inherent fallacy of Kobidi's starting premises'.

Weeks passed. No one rose to my defence. Philip Bousquet published another piece focused not on these invalid, regressive attacks on me but instead urging me to more 'temperate' discourse. 'I agree that Kobidi's theory is revolutionary. It may in fact, in the fullness of time, establish itself as the Theory of Everything. However, it is not illegitimate for Hess and Huslander to disagree with this assessment. It *is* improper for Kobidi to refer to these distinguished physicists as fascists (and bizarrely "Belgians". Neither of them is a citizen of Belgium. Hess is German and Huslander is an American). Kobidi is a brilliant physicist and, also, a close friend. With these letters he is steadily eviscerating whatever is left of his reputation.'

I continued sending out my letters, refuting point by point the absurd attacks by Hess, Huslander and their coterie, until one afternoon the Kinko's clerk, with a sad face, informed me that the credit card I handed to him had been declined. I was surprised when he refused to give back the plastic, claiming he was required to retain it. 'It's been reported stolen,' he said, pressing down on his lips, and then adding accusingly, 'Your name isn't Heidi, is it?'

As upsetting and unsettling as these attacks were to me at the time, they did not in themselves precipitate the activities that preceded my second confinement. That happened one morning when a two-page document was slipped under my door at Clark Hall. I did not rise to pick it up. I focused my attention on the papers I had gathered on my desk: copies of articles that touched on my last theorem. I had ignored the document beneath the door for over an hour when Anyanwu, who had recently started lounging in my office, rose, walked over and picked it up. He read out the title as he brought it to my desk: "'Copy Universe: Is Kobidi's Theorem Even Original?'"

I pulled it from his hands and hurriedly read through. The libellous screed, written by one of Huslander's postdocs, accused me of plagiarism – citing everything from obscure and irrelevant writings of some ancient Greeks to certain passages from

Einstein himself. There was a note, presumably penned by the individual who had slipped the papers under my door, informing me that the document was being circulated widely by email and on list servers.

I rose from my chair. Anyanwu could see what was in my eyes. 'Do not do it,' he said. 'There is nothing to gain. The world knows about you and about me now. Your work is recognised. Ignore them. Science will prove you are right.'

'You are advising caution? That I don't do anything?' I scoffed.

'I have warned you,' he said, turning away from me and raising to his eyes the issue of *Nature* in which I had published what would be my last theorem.

When I returned to my home later that afternoon, two police cars and an ambulance were in the driveway waiting for me. In a window on the second floor, I caught sight of my son, his mother behind him, moving away from the curtains.

On the day I left the lobby of the Richard H. Hutchings Psychiatric Hospital for the last time, my sister was waiting with my wife, in a black van, to pick me up. Sometime on the drive back to Ithaca, as I glanced out of the window at the passing trees, she placed a new issue of *Nature* on my lap. The bookmark intersected Professor Rayburn's letter to the journal: 'Professor Ezeani Kobidi's recent theorem is a mathematical proof. It is not a conjecture. It is not even a theory. It is a proof; an irrefutable logical conclusion. Kobidi does not theorise that entanglement produces space and time, he actually proves it.'

As I finished reading, my sister leaned in and held my hand.

17

DECOHERENCE TIME

The Uncertainty Principle

The more precisely we know one thing, the greater our ignorance of other things. The universe punishes the temerity of knowing it in one aspect by increasing ignorance in others. This is the real meaning of the uncertainty principle, one of the clear tenets of quantum physics.

In the year between my second confinement at the Hutchings Psychiatric Hospital and my death, my sister visited me often. She was in Ithaca at least once in each month. Some months, she came up more often. Once, for reasons that were somewhat unclear to me, my sister spent two weeks living in the house on Hanshaw Road with my wife, my son and myself.

She made this trip up the highways from Connecticut very early in the summer in which my natural life ended, several weeks before I died. Perhaps she felt, within herself, my upcoming death, long before the moment when she would see my body, bloated and constrained by an odd-fitting suit, in the white anteroom of a church's nave and weep. What we know is eclipsed by what we feel; she knew before her eyes produced the corpse that triggered her tears.

Her appearance at our door was a surprise to me, but I surmised, by the curt conversation between them, that this visit was something that had been discussed with my wife. Heidi held open the front door and ushered my sister into the house, a vacant look hanging over her face. This restrained look was now often on my wife's face, as if she was weighed down with the effort of forbearance.

In the two weeks she spent with us, my sister told me a lot about her life. She seemed eager, it appeared, to share with me anecdotes of hope and perseverance. A common theme was how she had overcome the challenges, isolation and difficulty that had confronted her when she first arrived in America. Perseverance had been needed to overcome the depredations of our father and, to a lesser extent, those of an unfaithful ex-husband. Once, sitting on our front porch, she said to me: 'As long as we are alive, there is hope. And when there is hope, life can eventually thrive.'

She was at this time also one of the few people still ready to sit and listen to an account of my last theory. She would sit patiently with me for hours while I explained its implications. Occasionally, she would ask precise, short questions. I would slow down and attempt to explain the point she had inquired about. She would listen to my description of Hess and Huslander's scurrilous allegations and push me, point by point, to refute them.

One afternoon on that last visit, just after the sun had reached its peak, my sister leaned over me and said: 'Ezeani, bia ka anyi puo. Come, let's go outside,' repeating her Igbo words in English, I presume for the benefit of Heidi who was seated on the couch at the end of the living room. Heidi's face turned up from the magazine on her lap, glanced vacantly in our direction and then slowly returned to the thick publication. I rose and walked, through the dining room, out through the screen door at the back of the house, my sister a pace ahead of me. I felt my wife was still watching.

The screen door shut behind us. We sat down at the wooden table. I looked out into the backyard, staring at the short heap of rocks. My sister spoke. Her words were almost always in Igbo, sometimes even when Heidi was in the same room. I tried to focus on those words.

They were about hate. The words rolled out in a staccato progression, different from the melodious articulation in which Igbo – in my mother's and her voice – usually flowed to my ears. I focused on the words. She was talking about our father.

'Hate is a useless feeling. It's not hate. It's pity that I feel,' were the words she was speaking.

'He was in America. He came to see me when I was in Syracuse. In the hospital,' I said.

'I know. Nnamdi told me.' My sister stopped speaking. I looked over at the rocks.

After a minute, I said: 'I don't think you killed our mother. I am sorry for those letters.'

'I know. You never thought I did, Ezeani,' she said. 'You were living in a madhouse with a madman. It was your way of coping.'

There was silence. Then, after a few moments, I could make out the sound of birds chirruping. My sister was looking at me strangely. Her face seemed sad. She got up. 'Are you OK?' she said as she hugged my head to her chest. She took the palm of her right hand and wiped tears from my face. 'Ezeani, you will be fine. It will be OK,' she said.

My sister stood there for a number of seconds holding on to me. Then she continued speaking in Igbo: 'That is what hurts me the most. When I think of it. It is how he treated you. It fills me with ọnụmà,' she said.

I heard all her Igbo words and understood them. Except one. 'What does ọnụmà mean?' I asked.

'I don't know that it's easily translatable to English,' my sister responded. 'Wrath and loathing wrapped into one, I suppose.' She sat back in the chair. We were quiet again.

'How are you and Heidi doing? The house seems different. I don't like it,' she said after a few moments had passed.

'We are fine,' I said. I did not know how to say, *She wants to leave*, without crying. I started to tap my hand on the table. After a few moments my sister placed her hand over mine and I stopped. 'She doesn't understand my physics. I don't think she likes me any more.'

'She looks after you, Ezeani. She is committed to you,' my sister said. 'Don't worry. It will get better. You will get better.'

'It doesn't feel like that,' I said. 'Everything feels different. Like ọnụmà has seized Heidi.'

The Standard Model of Particle Physics holds that there are four fundamental forces in our universe – the strong force, the weak force, the electromagnetic force and the gravitational force. The Standard Model is wrong; it contains an incomplete description. My final theorem established that there are two fundamental forces – love and strife; forces that bind things together and those that tear things apart.

My sister kept her hand wrapped around mine. She squeezed my hand. She didn't say anything. We sat in silence. 'Ezeani, listen to me,' she finally said, 'ọnụmà will not kill you. It is àrìrì that you need to protect yourself from.' I stared at her.

'Do you know what àrìrì means?' she asked. I shook my head. 'Grief? Deep sorrow? An infinite loop of heartache,' she said. I nodded.

'You know how you describe the Chi as the full potential. Everything a being can possibly be?' she asked. I nodded.

'Onye kwe Chi ya ekwe. Ezeani, don't ever limit your Chi. Never place it in the small room of despair,' she said.

Then the screen door creaked. 'Are you guys ready for lunch?' Heidi asked.

Of Laughter and Forgetting

My brother had a short glass of whiskey in his hand. He was leaning into the bar, whispering in the ear of the tall woman serving the drinks. She laughed. Then she shook her head. She turned to the large mirror at the back, briefly swiveling her head to look again at my brother. He smiled a small, tight smile.

Then he returned his attention to me. 'How are things in the sex department?' he asked.

I laughed. 'The department is closed. I can't get the key component to work any more,' I responded, and smiled.

'Perhaps the key component needs kinder, warmer persuasion,' my brother said, smiling back.

I pursed my lips. And I drank from my tall glass of orange juice.

'Look, Ezeani, I think you are on your way to the top of your game. You need to give yourself permission. If she isn't giving you what you need, then get it somewhere else.'

I did not respond. He put his arm around my shoulders and slid his bar stool closer to mine.

'It's your wife's fault. She opened up a line of attack. It is easy for them to dismiss the importance of your work. "Oh, he is mad!"' he said, waving his arms. 'And most people, of course, have no way of judging.'

'I have published my theorem. That's what's important. The idea is out of my head,' I said.

'Brother, you are wrong about that. Believe me. You can be naïve about these things. What is important is to fight her

287

nonsense. Some Caucasian will show up in a few years and say exactly the same things you just said and they will give him your Nobel. It's just bullshit,' he spat.

My brother had turned away from me. He was watching two tall women in short, black dresses walk into the lounge. His eyes followed them as they moved to a low table towards the rear.

'What if I am nothing?' I asked my brother. He turned his attention back to me.

'What do you mean?' he asked.

'What if I don't exist?'

'We are having a conversation, Ezeani. You and I. You exist.'

'What if you don't exist either?'

'Oh, I exist,' my brother said. 'You can count on that,' He smiled. His index finger was playing with the rim of the whiskey glass, but his eyes were fixed on the two women sitting at the end of the room. 'Excuse me,' he said, as he rose.

Soon, my brother beckoned me to join him. He introduced me to the women. They shook my hand and giggled. My brother called a waitress and ordered bottles of liquor, juices and soda beverages. When the waitress was pouring out my orange juice, my brother insisted she add a large measure of vodka. I drank the mixed drink. Then my brother filled up another glass and I drank some more. Soon, one of the women raised herself and sat on my brother's lap.

The night was dark and the street lights barely lit up the entrance to the lounge. I leaned against my brother. He was laughing.

'What is taking them so long?' he asked, I assumed rhetorically.

I leaned further into him.

'Did Obiageli tell you that our father tried to rape her?'

'Yes,' he responded. He answered quickly, barely looking at me, his eyes scanning up the street.

'Did you believe her?'

'I wouldn't put it past him,' he responded.

'Did you believe her?' I asked, again.

He dropped his arm. We started walking away from the lounge entrance. He placed one hand in his coat pocket.

'Yes, I suppose,' he responded. 'I did.' He looked down at the pavement. 'Who knows what really happened between them. Honestly, sometimes, I wonder if it's worse than Obiageli says.'

'When did she tell you?' I asked.

'When I first got to America. She told me about Ambrose too. These guys are just fucked up. When did she tell you?' he asked.

'When I first visited her in Connecticut with Heidi. She had letters I sent her from Ibadan.'

'She told you in front of Heidi?! Why would she do that?' Nnamdi said. I did not speak. I stopped and looked at my brother.

'Look, we need to be careful. It's information that can be used to hurt you and to hurt me, too. It shouldn't be in Heidi's hands,' he said. 'And you have to be aware that Obiageli isn't actually moved by these considerations. She just doesn't care. I mean at some level she acts like a saint. Do you know that she actually sends medicine to our father and that bastard Ambrose? She says the medicine is hard to find in Nigeria and without it they would die. But it's just fucked up. I know she doesn't give a shit. I don't know why she does it.'

I stared at my brother again. I was leaning into him and he was supporting some of my weight. We could both see the black SUV coming down the street, the sharp light breaking into the night.

My brother turned to me. 'Look, I can get you a room at the Statler and you can have either one of the girls. Take the first pick.'

'Really?' I asked and smiled.

'Yes, really, Ezeani. They are on the clock, my dear brother. A few hundred dollars for incalculable pleasure.'

My brother helped me into the back of the black SUV, then, at my request, he had the driver take me to my home on Hanshaw Road.

The Shadows of Other Selves

There was a storm the night my sister left our home on Hanshaw Road on that last visit, early in the summer of the year of my death. She left late in the afternoon, when the sun, like an orange spider, had started to crawl behind the edges of darkness. Before she walked out of the front door, she sat next to me in the living room and held my hand. She told me she would be back at the end of summer. She smiled and then hugged me. My wife walked out with her, escorting her to the rental car in our drive. They stood out there speaking for over forty-five minutes. I could see them, standing next to the car, through the windows. While they were speaking, Njoku came into the living room. He was holding a large piece of paper and a crayon. I picked him up and placed him on my lap.

'How old are you, Daddy?' Njoku asked me. I told him.

'Will you be alive when I am your age?' he asked.

'I don't know,' I responded.

'I want you to be alive when I am your age,' he said. 'I will take you in an ambulance to the doctor so you won't be sick and die.'

I smiled. He hugged me. Then he showed me the large piece of paper. 'I am doing math,' he declared. I smiled. 'Give me a math question,' he requested. He climbed down from my lap and positioned himself on the rug, crayon in hand, ready to be tested.

'Forty-eight minus sixty-four,' I said.

He mouthed the numbers. 'Hey Daddy, sixty-four is more than forty-eight,' he said, as if he had caught me at a little trick. 'Do you mean sixty-four minus forty-eight?'

'No, my son,' I said. And then I taught him the non-intuitive logic of negative numbers.

I was asleep in the attic when the storm came. I could hear the rain on the roof. I turned in the bed and fell asleep again. That night I had a dream that was the most lucid and clear that I had ever had in my life. I only realised I was dreaming when I woke and found myself still in the small-frame bed in the attic of our home on Hanshaw Road.

In the dream I was wandering about in a savage downpour. I had a simple cloth wrapped around my waist. It took me a moment to realise that I was in the village, in Umudim, the first place I ever lived. The rain came down in a fierce rage, the drops pelting my skin. It was night. My path was only lit by the moon which would sometimes be swallowed by clouds. I didn't know where I was going or what I was looking for, but I had a strong sense that I was searching for something and heading somewhere. There was a path in front of me, and the rain was making the earth wet. I was not wearing shoes. I could feel mud between my toes. A fissure of lightning rendered the darkness. The large clouds moved away from the moon, and as if it had risen from the earth, a large lodge emerged in the path directly ahead. The lodge was held up by wide wooden beams and had a thatched roof. Wooden statutes and some other carvings were gathered on each side of the door. When I climbed up to the entrance, the wooden door swung open to let me in. The room was filled with the scent of smoke. Two women were seated in front of a large fire, their legs straight out in front of them. The one furthest from me rose to turn a wooden ladle in the large pot that stood on the fire. There were things hanging from the rafters, smoked fish and meats and prawns. The woman stepped over what looked like a log as she sat down again.

I looked up at the fire. Anyanwu was mounted directly above the flames, on a white and pink stool tied to a large beam that ran the length of the lodge. He had seen me. I raised my arm to salute him and greeted him formally. He couldn't raise his arms to reciprocate. They were tied to his sides. As he opened his mouth, flames instead of words poured out. The women turned to me. I immediately recognised the one closest to me as my mother. 'Nwam, please come and sit down,' she said, motioning with her arm. My mother turned to the other woman and said: 'Our mother, Ani, Empress Goddess of the Earth, my son has come.' She was smiling. Ani grunted. As I got to my mother, she reached out her hand and placed it on my head. 'Welcome, my son,' she said. Then she held my face between her hands and stared.

Ani looked over at me. Then she jutted her head towards Anyanwu. 'Do you know him?'

'I know him,' I said.

'Who is he?' she asked.

Anyanwu tried to speak but flames just spat out, roaring in the darkness.

'I asked you a question,' the Earth Goddess Ani said, her voice hard. 'Who is he?'

'Anyanwu, the Sun God.'

Ani the Earth Goddess cackled with laughter, the mirth creasing the lines of her worn face.

'My son, let me get you something to eat. Yam pepper soup

with dried fish. It is one of your favourites.' My mother stood up with a bowl in her hand and moved towards the large pot on the fire. As she stepped over the log, I saw that it was not a log at all but my father, encircled by the luminescent body of a large python. The python's head moved as my mother stepped over it and a forked tongue flicked out of its mouth. My mother returned with the bowl of hot pepper soup and placed it between my hands. I lifted it to my mouth and drank. The warm soup filled me. When I removed the bowl from my lips, my mother was staring at me with a smile on her face. 'Nwam,' she said. I smiled back at her.

Converging Worldlines

About a week before the end of my natural life, four days before I died, I walked down the stairs of our home at approximately 3.02 a.m. The only light that was still on in the room was a lamp on a table beside an armchair. It threw light in a tight circumference onto the side table. I picked up the receiver of the telephone on it and dialled my brother. The phone rang for a long time. It was not picked up. I dialled again. Immediately it rang the second time, he picked up. There was a lot of noise in the background. It sounded like a party. He was speaking but I couldn't make out his words.

Then finally the noise subsided. There was a rhythmic thud in the background like the noise had been imprisoned in a bag.

'Are you at a party? Am I disturbing you?' I asked.

'No, no, go ahead,' he said.

'They are trying to kill me.'

'What?'

'They are trying to kill me. I can sense it. I don't know what they are planning.'

'Who?' he asked. 'Who is trying to kill you?'

'I can't say it over the phone.' I quickly hung up.

The next morning my wife called me downstairs to the house phone. When I placed the receiver against my ear I heard my sister's voice.

'Ezeani, kedu? How are you feeling?' she asked.

'I am fine,' I responded.

'I am worried about you,' she said. 'Nnamdi said you called last night and said someone is trying to kill you.'

'No one is trying to kill me,' I responded.

'I can come up to see you tomorrow,' my sister said.

'I am getting better. Really I am. There is no need for you to come up. I am planning on taking Njoku out on Skaneateles Lake the day after tomorrow. We did it last year. We are going up with his grandfather. I am looking forward to it.'

'Are you sure?'

'Yes, I really enjoy going out on the lake with him. He is such an interesting boy. Do you know what he said to me the other day?' I started talking for about thirty-five minutes on things my son had said to me. My sister stayed on the line listening. Sometimes she commented. By the time she got off the phone she was laughing. I was laughing too.

When I hung up, I saw that Heidi had sometime during the conversation come back into the room and seated herself on the couch at the far end.

'Ezeani, what is going on with you? Why are you doing these things? Why won't you let me help you?' Her face was drawn.

'What happened? What did I do?'

'Why would you go to the police station and tell them that

you have information about a church arson that occurred almost five years ago?'

I had no recollection of having done any such thing. I stared at my wife.

'You are reckless. These statements jeopardise whatever is left of your reputation. You are destroying the work we have built together. Let me help you,' she said.

'With what?' I turned. Heidi just stared at me, her eyes tight and grey as if ọnụmà had seized her.

The Man Died

On the morning of my death, I planned to take my son out on the lake at Skaneateles. He was getting dressed in his room with the assistance of his nanny. In our bedroom, I had just pulled on a black T-shirt over a pair of red shorts. I picked out my white boat shoes and put them on. My father-in-law would be at the house in less than an hour. I headed downstairs to wait for my son in the living room, but as I reached the landing, I heard my name being called from the top of the stairs.

'Please come and see something,' Anyanwu said. He was dressed in the white, black and red checkered cloth, the ends magisterially thrown over his right shoulder. He was wearing the red cap with a large, gleaming eagle father placed erect behind his ear. His beard was trim. He seemed to have made an effort to convey something of his majesty. His chest was oiled, and the muscles rippled. I didn't have time for whatever he had in mind. My mind was focused on the excursion with my son. But he insisted. I was irritated as I walked up the stairs.

Immediately I entered the attic room, Anyanwu offered me

the desk chair. 'Please sit down,' he said. 'I have something important to tell you.'

'I have somewhere to go,' I said. 'I don't have time for this now.'

'It won't take long. After this, I won't be disturbing you.'

I sat down. 'What do you want to tell me?' I asked.

'I have something for you to drink. I will tell you after you drink.'

He pulled out a blue plastic cup with a dark fluid in its bottom quarter and handed it to me.

'Take and drink,' he said.

'No, I am not drinking today,' I said.

'Drink. It is good,' he said. I looked at him. Then I threw the liquid into the back of my throat. I had expected the burn of alcohol. It tasted like nothing, perhaps slightly sour.

Anyanwu was looking at me strangely. There was a smile on his face.

'What is it you want to tell me?' I asked.

'The thing I have given you to drink will kill you. Your wife said it. The way you are behaving will ruin everything. The letters you are sending to everyone. The reputation and respect that I have earned will be destroyed. People know me now. There are people starting to worship me again. All this is going to be eviscerated by what you are doing.'

'You are killing me?!' I screamed in disbelief.

'Yes. You are the sacrifice. It is Jesus that actually gave me the idea. You have to be killed so you can rise again. You have to be sacrificed. You need to be a martyr. Everyone will know that I, Anyanwu, sacrificed you to save the world. They will worship us both!'

It started to become clear to me that I was in a bad situation. I could sense that whatever Anyanwu had given me to drink was acting quickly. I willed myself to get up and walk unsteadily to the desk.

'What are you doing? What are you looking for?' Anyanwu asked.

'Is there an antidote?' I asked.

'No,' Anyanwu said and smiled. 'There is no antidote. You have to die. That's the only way it will work.'

'Have you considered that you might have miscalculated?' I asked. 'Maybe Jesus is fooling you? What if everyone forgets me . . . and you?'

There was a sudden alarm in Anyanwu's eyes.

'Where is the antidote?' I asked.

'There is no antidote,' he said. 'You will come back.' The alarm was gone from his eyes. 'Don't worry. I believe in you. I believe in your physics. We are entangled. We are always with each other.'

I sat down on the floor and pulled my journal to me. I opened it and placed a pen against the paper. It was important that I use these moments well. It was unclear to me how much longer I had to live.

'What are you writing?' Anyanwu asked. I ignored him.

Anyanwu sat down beside me and stretched his legs out, as the life ebbed out of my being.

We sat there for a few minutes. I could no longer see shapes clearly, like I was looking through a rag.

Before my time stopped, I heard Heidi knocking gently and then with increasing frequency. Anyanwu stood and stumbled to the door. He opened it and ran out past my wife. I collapsed to the floor. I was dead before Heidi walked into the room.

18

THE BURIAL OF THE DEAD

In Negative Time

The night is cool. I can feel my mother's heat through the wet cloth. Her warmth passes through soaked cotton and the drenched knotted fabric that secures my body to her back. The rain falls in bitter sheets, striking at my mother and then at me. I open my eyes. I can make out the dim lights of the white building, the light waves refracted through the moving water drops so they smear my vision to a magical blur. Soon, we are close enough and I can read the sign at the entrance: **Local Government Rural Clinic, Umudim.**

The nurse takes me from my mother. Her hand is rough. She carries me like a sack. 'He is sick again?' she asks, shaking her head. 'What is wrong with this child?' My mother is weeping. I can see this clearly, now that we are sheltered from the storm. When she wipes her face, the tears reappear almost instantly. The nurse is saying something to my mother as she places me on a table. The light in the room is strong and bright, blinding my eyes.

We have been moved to a different room. It is just the two of us. My mother and I. It is a small room. My mother's eyes are fixed on me. She is smiling, but only slightly. She plays with the hair on my head, then she takes my hands in hers. Outside,

I can hear the voice of the rain. The only light is a dim hurricane lantern, the flame around the wick fluctuating in an uneven pulse.

'Nwam,' my mother says, 'you will not die.' I stare back at her. 'Do you hear me, my son? You will not die.' She is still smiling. The light falls weakly on her face, leaving blurry shadows that drop to the floor. Suddenly, my mother starts crying. She pulls her right hand to her face. Her chest heaves. She shakes her head, over and over.

'Nne, I will not die,' I say. Immediately I speak, her tears stop. The look on her face seems sad and sceptical.

'I promise you my mother, I won't die,' I say, lifting my head slightly. My mother hugs me to her. My nose fills with her scent.

'May it be so, my son. May Ani help us!' my mother says. Soon the rain stops. When the dawn starts to slip in through the window, my mother is asleep on the floor beside the short bed. She moans something I can't quite hear. Then she waves her arm, as if she is shooing a nuisance away. Soon, I fall asleep.

In Ordinary Time

Like eggs, five red mounds of heaped earth were arranged in a line around the grave. The mid-afternoon sky was dark with turbulent clouds. My father was on one side of me, and my brother on the other. We were all wearing black. The priests, in purple and gold chasubles, with cream soles hanging from their necks, were speaking quickly. 'We commend your soul, our sister in Christ, Agnes Nwanyibuife Kobidi, to almighty God, and entrust you to your Creator,' their leader said. Just as he finished, the rains' violence was unleashed by the dark clouds. Everyone ran off to the shelter of the building's awning. I stood beside my mother's coffin. The rain pelted and then drenched me. The red earth ran with the falling water into the grave.

When the short, violent storm ended, the sun emerged from behind the lightening clouds with a fierce burn. It heated my head and the back of my neck. Then the gravediggers arrived and placed my mother's body into the damp earth. The red soil stained their feet.

In Imaginary Time

A negative number divided by another negative number is a positive number. An imaginary number divided by another imaginary number is a real number. Imaginary time divided by imaginary time is real time.

The flies buzzing around my face were starting to make me uncomfortable. This was not the way I thought my body would be treated. It didn't even seem warm enough for flies. Yet, despite the cool air and the formaldehyde, I smelt the deep stink of decay. My body was filling up with putrid gases, the unrelenting process of decomposition. The attendant waved the flies away, then he sprayed something in the air. He took a few breaths and then he walked to the door and opened it. My sister walked in. She stood over me. Her face was tight. There were already tears in her eyes. I wanted her to leave. I wanted to spare her nose the whiff of decay.

She stayed a long time. Standing over me. And weeping. Then she touched my face, and turned and ran out of the room.

The attendant closed the door behind my sister. I was surprised. I thought he had left. I hadn't noticed him. Quickly, in a regulated, almost regimental manner, he walked over to my corpse, closed the coffin's lid, then wheeled the platform on which the coffin rested through another set of doors into the Church of Immaculate Conception.

One cannot understand time until one understands life. Nothing that connects with another – the howl of sound, a straggling ant, a frightened boy, a spirit, a man, a God – is outside of time. Nothing means anything except in connection with something else. Time is where we count our connections; how one changes in relation to the other. This is what time is. This, also, is what life is.

My son is in the first pew, he is distracted, playing with his mother's hair. My wife's father sits beside her. Beside my father-in-law sits my own father. And beside him my brother. They are all wearing black. My wife is also wearing a black hat. She pulls my son to her as the trolley with my coffin rolls in. There is a line of priests in purple and gold vestments behind the altar. The leader of the priests starts speaking.

My son is not paying attention. He has started playing with my wife's hat. My wife gently tugs at his arm. I want to say something to my son. I want to tell him what I have told you. I cannot open my mouth but still, I will myself to speak.

My son stops his play. He looks at the altar, then at the priests. He turns from them and walks into the aisle. The priest stops speaking. My son starts running down the aisle towards the doors. He is yelling something. I can't hear what he is saying. My wife stands to run after him.

Nothing means anything except it is connected with something else.

The Comfort of Distant Stars

by Ezeani Kobidi
Class 4B, Staff School, Ibadan

Darkness and star-speckled space
Distant beings that blink and smile
Cool lights never blind or scar

An adjacent sun burns crude curses
Growls fierce like the pacing leopard
Fevered giant dithering and vile
Dark person clouded by light

Darkened space and distant stars
Ancient fiends that slink and slide
Feeble lights do not burn or blight

The sun teases us with warmth
Mocks the giant and the dwarf
Blinds the terrible leopard with sight
Succouring murk devoured in spite

Our sorrows embrace this dawning light
Distant suns do not shade our smile
Senile specks always far and slight

ACKNOWLEDGEMENTS

My gratitude to my mother, Dr Rose Nkeonyere Echeruo, for reading this book in draft and for a lifetime of love; and to my sister Ijeoma Echeruo, for her incisive observations on the text.

I would like to thank Iguwo Ukwu, Muhtar Bakare and Dayo Ogunyemi for reading the manuscript and providing comments and encouragement, and to acknowledge the great generosity of my friends Ike Anya and Chika Unigwe.

Over several years, I have relied on sources too numerous to list for information and insight into the world of mathematics and theoretical physics. However, I am particularly indebted to the book *Quantum Reality* by Nick Herbert and the article 'Relativity and the Problem of Space' by Albert Einstein.

Kobidi's fictional achievements are based on the contemporary work of numerous physicists, sometimes pushed back in time. In particular, his 1990s fictional advance in combinatorial physics is based on seminal work in 2000 by Alain Connes and Dirk Kreimer demonstrating that the renormalisation of Feynman diagrams can be described by a Hopf algebra and related advances in permutation in the following decades, including the mid-2000s work of Ruth Britto, Freddy Cachazo, Bo Feng, and Edward Witten setting out equations that condensed hundreds of pages of Feynman calculations to single lines. Kobidi's last theorem (and its underlying idea that entanglement creates space and time) is a fictional extrapolation of the holographic principle of Nobel Prize-winning physicist Gerard 't Hooft and physicist Leonard Susskind, and AdS/CFT correspondence, initially identified by physicist Juan Maldacena. It is important to state that the idea underlying Kobidi's fictional theorem has not been proven, especially as it relates to emergent time, and the subject remains an active area of research by several physicists including Mark Van Raamsdonk, who has written on building up spacetime with quantum entanglement.

For her insight, intelligence and attention, I am indebted to my editor, Ellah Wakatama; and to the team at Canongate for their faith and dedication to this book, especially Vicki Rutherford, Leila Cruickshank, Brodie McKenzie, Rali Chorbadzhiyska, Lucy Zhou, Melissa Tombere, Phyllis Armstrong, Alice Shortland, Gabrielle Chant and Francis Bickmore.

And for the time and space to create, I am grateful to, *ndi ji obi'm*, Indira, my wonderful wife; Diala, Ulari and Ziora, joyful reminders of childhood's precocious light.